Allen Lane

The Penguin Book of the Bicycle
Roderick Watson and Martin Gray

ALLEN LANE
Penguin Books Ltd
17 Grosvenor Gardens
London SW1W OBD

First published 1978
Published simultaneously in paperback by Penguin Books

ISBN *0 7139 1099 2*

Set in Monophoto Ehrhardt

Printed in Great Britain by
Fletcher & Son Ltd, Norwich

To Katie and Celia
and for Christopher, Laura,
Joanna and Nicholas

Contents

Preface

An entire volume could have been written on almost every one of the chapters in this book. We hope that our approach will stimulate interest in what the bicycle has to offer and enhance the entertaining and human dimension of the machine the French have called 'little queen'.

M.G., R.W.
January 1977

Acknowledgements

We would like to thank the following organizations and individuals who made their time and experience available to us with such unfailing courtesy: the Reynolds Tube Co. Ltd, Raleigh Cycle Industries, Birmetals Ltd, the Dawes Cycle Co. and Dales Cycles, Glasgow; Gerald O'Donovan, Ken Evans, Dave Moulton, Richard Phipps and Richard Holt. We would like to express our indebtedness to those who have written and researched on the subject before us, especially to the editors and staff of *Cycling*, and also to *Cycletouring, International Cycle Sport, Bike World* and *Bicycling*. We are obliged to the library staff at the Cyclists' Touring Club, the Mitchell Library, Glasgow, Stirling University and the Imperial War Museum.

Finally a special thank-you must go to the people, too many to name, who have helped us and listened to us so patiently; to the cycle shops who put up with our questions and gossiped so pleasantly; and to our friends in the mid-Scotland cycling clubs.

Publishers' Acknowledgements

Grateful acknowledgement is made to the publishers for permission to quote from the following works:

H. G. Wells, *The Wheels of Chance*, J. M. Dent & Sons
Jerome K. Jerome, *Three Men on the Bummel*, J. M. Dent & Sons
Terence Bendixson, *Instead of Cars*, Temple Smith
Daniel Behrman, *The Man Who Loved Bicycles*, Harper Magazine Press, New York
Robert M. Pirsig, *Zen and the Art of Motorcycle Maintenance*, The Bodley Head
Ivan Illich, *Tools for Conviviality*, Fontana/Collins
Ivan Illich, *Energy and Equity*, Calder & Boyars
Samuel Beckett, *Molloy*, Calder & Boyars
The Velocipede: Its History and How to Use It, Cyclists' Touring Club
Lewis Mumford, *Art and Technics*, Columbia University Press, New York
S. S. Wilson, 'Bicycle Technology', *Scientific American* (March 1973)
R. E. Williams, 'De Motu Urbanorum', *British Medical Journal* (4 October 1975)
Hugh Kenner, *Samuel Beckett: A Critical Study*, John Calder
Flann O'Brien, *The Third Policeman*, Hart-Davis MacGibbon
Flora Thompson, *Lark Rise to Candleford*, Oxford University Press
Richard Hoggart, *The Uses of Literacy*, Chatto & Windus
Ernest Hemingway, *A Moveable Feast*, Jonathan Cape
Carl Bernstein and Bob Woodward, *All the President's Men*, Martin Secker & Warburg

1 The Bicycle in Fashion Again: Some History and a Modest Manifesto

A velocipede riding school in New York, 1869.

In the early part of January, 1869, I was at Spencer's Gymnasium in Old Street, St Luke's, when a foreign-looking packing case was brought in . . . A slender young man, whom I soon came to know as Mr Turner of Paris, followed the packing-case and superintended its opening; the gymnasium was cleared, Mr Turner took off his coat, grasped the handles of the machine, and with a short run, to my intense surprise, vaulted on to it, and, putting his feet on the treadles, made the circuit of the room. We were some half-dozen spectators, and I shall never forget our astonishment at the sight of Mr Turner whirling himself round the room, sitting on a bar above a pair of wheels in a line that ought, as we innocently supposed, to fall down immediately he jumped off the ground.

Ixion – A Journal of Velocipeding, Athletics and Aerostatics (1875)

Such was the wonder caused by the first French velocipede brought to London. The slender Mr Turner was Paris agent for the Coventry Sewing Machine Company. Within a few months he had persuaded his uncle, the manager of the company, to manufacture these miraculous new machines, and the Coventry cycle industry had begun. The foreman of Turner's Company, under its new name, Coventry Machinists, was a certain James Starley. By the 1870s he had greatly modified the crude velocipede and pioneered

J. K. Starley on the Rover.

the high-wheeler. By 1885 his nephew, John Kemp Starley, had produced the elegant 'Rover', with its diamond-shaped frame. The modern bicycle was born.

In Britain today the bicycle is no longer associated with wonder or delight. Yet it has given to the language that excellent word 'free-wheeling', pleasant in all its different meanings. It led directly to the invention of the ball bearing, the tangent spoked wheel and the pneumatic tyre, innovations which, according to Ivan Illich, can

Italian Futurist Umberto Boccioni made these studies for his 'Dynamism of a Cyclist' in 1913.

only be compared with three other events in the history of transportation: the discovery of the wheel itself; the simultaneous invention of the stirrup, shoulder harness and horseshoe, which revolutionized the agricultural economy of medieval Europe; and the building, in the fifteenth century, of the first real ocean-going vessels, which so greatly advanced world trade.

As you bang into a suddenly opened car door, you may care to reflect on Illich's assessment and to see irony in the fact that your bicycle so unequivocally paved the way for the mass production of the motorcar. Many of the famous names – Ford, Hillman, Morris, Peugeot, Opel are but a few – first acquired a reputation in the bicycle industry. Garage firms, if they date back any time at all, usually turn out to have started by mending and selling the Ordinary, or penny-farthing. The bicycle was for a long time associated with technical advances and forward thinking. The Wright brothers were a cycle firm, and the Italian Futurists – iconoclastic heralds of the age of electricity – championed the racing cycle as an image of grace, speed and power. Yet for many years since it has been a forgotten drudge, the Cinderella of technology.

In post-war Britain pedal-driven transport had an aura of shabbiness, failure, and threadbare unfashionableness. This image has persisted in the most unlikely places. Thus when the literary critic A. Alvarez sought to commend a more powerful kind of poetry than had been written by English poets of the nineteen-fifties, it was cycling which crystallized his dislike for the politeness and orderliness of their work. The essay was written in 1962 and was called 'Beyond the Gentility Principle':

> The pieties of the Movement . . . are summed up at the beginning of Philip Larkin's 'Church-going':
>
> *Hatless, I take off*
> *My cycle clips in awkward reverence.*
>
> This, in concentrated form, is the image of the post-war Welfare State Englishman: shabby and not concerned with his appearance; poor – he has a bike not a car; gauche but full of agnostic piety; underfed, underpaid, overtaxed, hopeless, bored, wry.

The reverse of this picture, the positive of Alvarez's negative, so to speak, is not likely to be a cyclist. His image permeates our consumer culture as he beams at us from the Sunday supplements and the television set. He is smart, and very concerned with his appearance; rich – he *must* have a car, not a bike; assured, but full of humanist sympathies; overfed, overpaid, optimistic, interested in everything, complacent. For such a person the bicycle had a stigma of poverty or eccentricity, suitable for the new genteel poor, pre-adolescents, or the working man who failed the Joe Lampton climb to that room at the top.

It was not always so. From about 1890 till the turn of the century cycling was highly fashionable and hugely popular. The writers of

Lunchtime at a dockyard, summer 1939.

women's society pages would report on Lady So-and-So's cycling outfit, in colours which matched the livery of her machine; and the editors of the *Cyclist* (Vol. XXIV, no. 1,184) could print the following accolade in all seriousness:

'Long Live the King!'
That brief and strenuous prayer will assuredly rise at once to the lips of the wheel folk, for is not Edward VII, King of Great Britain and Ireland, of the Britons beyond the Seas, and Emperor of India, himself a cyclist?

A king, after all, is only a man and even the humblest subject may pedal as fast – if he dares to risk such lese-majesty. Such was H. G. Wells's Hoopdriver in *The Wheels of Chance* – a draper's assistant and one of thousands who were liberated from the cities by the advent of the mass-produced safety bicycle.

The freshness of dew was in the air; dew or the relics of an overnight shower glittered on the leaves and grass. Hoopdriver had breakfasted early by Mrs Gunn's complaisance. He wheeled

A Royal Pastime.

his machine up Putney Hill, and his heart sang within him. Half-way up, a dissipated-looking black cat rushed home across the road and vanished under a gate. All the big red-brick houses behind the variegated shrubs and trees had their blinds down still, and he would not have changed places with a soul in any one of them for a hundred pounds.

He had on his new brown cycling suit – a handsome Norfolk jacket thing for 30s. – and his legs – those martyr legs – were more than consoled by thick chequered stockings, 'thin in the foot, thick in the leg', for all they had endured. A neat packet of American cloth behind the saddle contained his change of raiment, and the bell, and the handle-bar, and the hubs and lamp, albeit a trifle freckled by wear, glittered blindingly in the rising sunlight. And at the top of the hill, after only one unsuccessful attempt, which, somehow, terminated on the green, Hoopdriver mounted, and, with a dignified curvature of path, began his great Cycling Tour along the Southern Coast.

There is only one phrase to describe his course at this stage, and that is – voluptuous curves. He did not ride fast, he did not ride straight, an exacting critic might say he did not ride well – but he rode generously, opulently, using the whole road and even nibbling at the footpath. The excitement never flagged . . .

. . . At the crest of the hill he put his feet upon the foot-rests, and now riding moderately straight, went, with a palpitating brake, down that excellent descent. A new delight was in his eyes, quite over and above the pleasure of rushing through the keen, sweet, morning air. He reached out his thumb and twanged his bell out of sheer happiness.

'He's a bloomin' dook – he is!' said Mr Hoopdriver to himself, in a soft undertone, as he went soaring down the hill, and again: 'He's a bloomin' dook!' He opened his mouth in a silent laugh.

As the issue numbers of the *Cyclist* indicate, it was a boom period for cycling, and newly fledged riders were everywhere with a positive explosion of makes and model improvements, patents, ingenious devices and accessories of all kinds. Jerome K. Jerome had obviously suffered some of these, as he recounts in *Three Men on the Bummel*:

'Then there are saddles,' I went on . . . 'Can you think of any saddle ever advertised that you have *not* tried?'

He said: 'It has been an idea of mine that the right saddle is to be found.'

I said: 'You give up that idea; this is an imperfect world of joy and sorrow mingled. There may be a better land where bicycle saddles are made out of rainbow, stuffed with cloud; in this world the simplest thing is to get used to something hard. There was that saddle you bought in Birmingham; it was divided in the middle, and looked like a pair of kidneys.'

He said: 'You mean that one constructed on anatomical principles.'

'Very likely,' I replied. 'The box you bought it in had a picture on the cover, representing a sitting skeleton – or rather that part of a skeleton which does sit.'

He said: 'It was quite correct; it showed you the true position of the –'

I said: 'We will not go into details; the picture always seemed to me indelicate.'

He said: 'Medically speaking, it was right.'

'Possibly,' I said, 'for a man who rode in nothing but his bones. I only know that I tried it myself, and that to a man who wore flesh it was agony. Every time you went over a stone or a rut it nipped you; it was like riding on an irritable lobster. You rode that for a month.'

'I thought it only right to give it a fair trial,' he answered.

If Hoopdriver's golden age exists anywhere today it might be found in the pages of the club magazine *Cycletouring*. The Cyclists' Touring Club was founded in 1878 and its fluctuating membership since then reflects many a change in fashion and prosperity. The Club is a valuable organization and *Cycletouring* takes an active role in putting the cyclist's case in a world of motorcars. Often it has its own kind of Hoopdriverish optimism: 'The nod, the wave, the smile, the greeting between cyclists as they pass along the road is something that should be fostered – a worthwhile reminder, amid the hurly-burly of mechanized traffic, that people matter.'

You might be forgiven for feeling at times that the world of *Cycletouring* is strangely fabulous, as if another Britain existed filled with country lanes, thatched cottages, fields and rolling hills, populated by a race of true Britons, gentle, polite and cycle-clipped. Bicycling becomes a kind of escape, and the modern world is scathingly mentioned from time to time, in the full knowledge of every reader's disapproval. You hope desperately that this dream world of country lanes *does* exist. But you suspect that somehow it is a matter of faith, and that most of us are excluded from the CTC congregation. *Cycletouring* is a delightful magazine. It is like a smaller, more modest *Country Life*, without the grandiose advertisements for antiques and houses, but with a little bicycle leaning

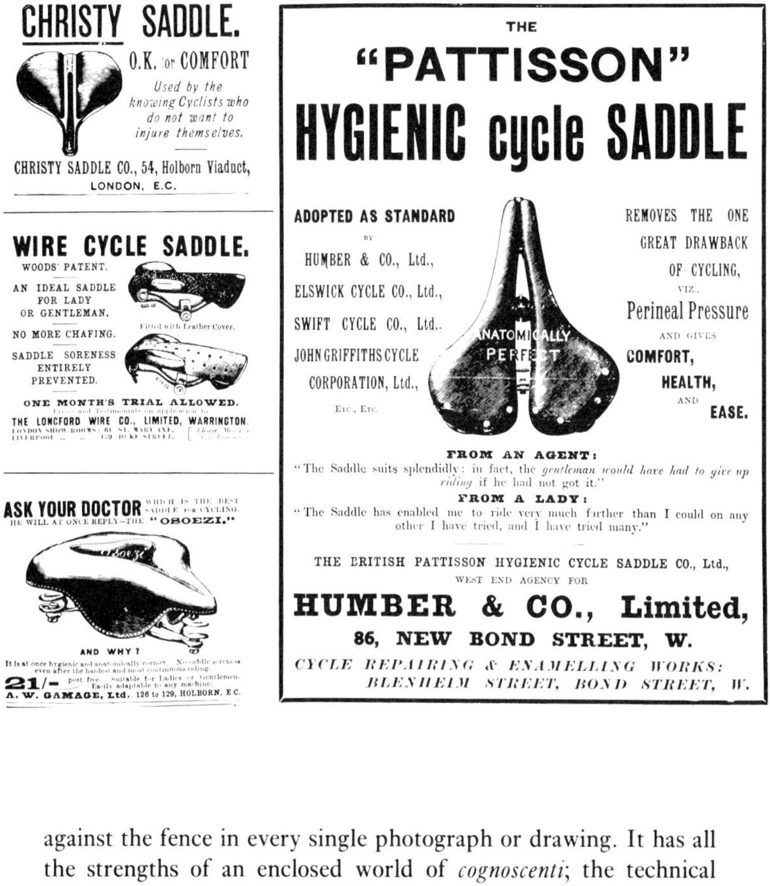

against the fence in every single photograph or drawing. It has all the strengths of an enclosed world of *cognoscenti*; the technical information is meticulous and fascinating, while individual members offer their damaged sidecar to any interested party or ask for the owner to apply for a pair of gloves discovered after a Club Birthday Ride. This intimacy is pleasant and worthwhile but not likely to change the image of cycling in Britain, and the CTC may yet have to learn to be more ruthless in the business of protecting its members.

Recent trends do indicate that the bicycle lobby is becoming more actively vociferous but, sadly, for most of the adult population of Britain the cycle still has the status of a toy. And even then it is liable to neglect, for the roads are too dangerous for children. As Terence

Bendixson points out in *Instead of Cars,* we have learned to accept without question a state of affairs which is, at the very least, bizarre:

The reality is illustrated by the TV road safety advertisement that shows a child being smashed by a car followed by a hammer

Cycle touring in 1929 as seen by the inimitable Frank Patterson, whose work was first published in the weekly *Cycling* in 1893.

hitting a peach. How did we ever let our peaches get mixed up with millions of swinging hammers? And what sort of pride can we take in having calculatedly to build up fear in our children so that they can survive in the mechanical jungle we have created.

Bendixson's is one among several recently published books which note that in planning and altering our cities with the motorcar in mind we have made them increasingly unattractive and dangerous for people to live in.

Apart from children, who uses the bicycle? Sir Colin Buchanan's examination of Newbury in *Traffic in Towns*, published in 1963, showed that 17 per cent of trips to work were made by bicycle, compared with 36 per cent on foot or using public transport. He recommended that £4½ million be spent on urban motorways, but he decided that it would have been 'very expensive, and probably impracticable, to build a completely separate system of tracks for bicycles'. His suggestions typify the official response to the cyclist's needs during the last two decades. Glasgow, for example, has been

Playing roller-skate hockey in a New York street.

sliced apart by an obtrusive and extremely expensive trail of urban motorways; but Glasgow has the lowest number of car owners *per capita* of any city in Britain. Such planning makes no sense unless it goes hand in hand with universal car ownership, and conventional assumptions about this are often misleading. It is important to remember that, even if *every* family has a car, the man who takes it to work leaves his wife, children and possibly parents effectively stranded. And yet, all too often, the town they are left to walk in has been radically restructured to suit the needs of that vehicle standing silently in the company car park.

A new stretch of urban motorway was opened in Glasgow in 1975 at a cost of £7 million. A footbridge crossing it in a less well-off area had to be fully enclosed to protect the cars from the effects of stones and the attentions of the local children. Maybe they were trying to tell us something.

Parked bicycles in Peking. There are almost no private cars and the two-wheeler is the main form of personal transport.

It is surprising that the bicycle is not more widely used than it is when we consider how freely it moves in traffic and that the average car-journey in town is only four and a half miles. Furthermore, the new small-wheel and post-Moulton designs can cope with as much shopping as can normally be carried by hand. It is now quite clear that the famous twenty-three miles of cycleways in Stevenage did lead originally to a significant increase in cycling, both for pleasure and as a means of getting to work. But it is also clear that the urban motorway, with its humps, flyovers and underpasses, makes life for the cyclist even more difficult, if not impossible. Given that some towns have geographies that are kind to cyclists, it is a lamentable

Roundabout and cycleways at Stevenage – a safer deal for pedestrians too.

shame that more councils do not make some kind of effort to foster the bicycle as a commuter's vehicle. It is cheap to run, and relatively cheap to cater for: one square yard of cycleway costs a sixth of the amount needed for a square yard of carriageway, and a cycleway has ten times the carrying capacity of a carriageway. About twelve bicycles can be parked in the space taken up by an average car. And as for safety, Stevenage estimated that cycleways saved about £28,000 per annum in hospital expenses. Vasteras, in Sweden, where an elaborate system of cycleways led to 25 per cent of all journeys being made by bicycle, reckoned to have saved £100,000 every year through a reduction in accidents because of less traffic congestion.

Daniel Behrman is an American writer in Paris who enthuses about life on two wheels. In *The Man Who Loved Bicycles* he condemns the private car for the pollution it causes and points to another kind of danger it creates:

> But don't let exhaust fool you, it's only a smoke screen. Run an automobile on steam, electricity, sunshine, or the morning dew, it'll still get you. Put on bumpers of eiderdown, bring back the man on horseback waving a red flag ahead of every motorcar (why did they ever take him away?), the automobile will still be lethal.
>
> For that deadly Mustang–Cougar–Jaguar–Tiger GT you take by the tail is really nothing but a wheelchair. The difference is that most wheelchairs give the patient a chance to push the wheels. Hardly a wheel to push or turn in the deadly Mustang–Cougar–Tiger etc., it demands no more effort than the paraplegic's eyelid flick that flips the pages of an electrically operated book. A buddy of mine who had a foot nerve severed in the infantry got a priority in 1945 for an Olds Hydramatic; everybody has a priority today. Everybody is a paraplegic, we have superpower for infra-people . . . no human beings until now have ever commanded so much artificial energy while using so little of their own.
>
> The calories pour into the gut; the gasoline that goes into the tank is converted into motion, but not the calories in the gut.
>
> . . . For a while, we can do anything, smoke, drive, eat, drink. Not forever, though; the furnace stops drawing, we keep shovelling the stuff in but it doesn't go away. Cut down on sweets, try cyclamates instead; no, back to saccharine again; anything that will save the sweet sensation of fuel going down the gut without forcing us to convert it to work. Try the drinking man's diet, try the driving man's diet.

> That is the true physical pollution of the car. The maimed and the dead are the tip of the iceberg, the gassed and the poisoned are but part of the picture. The real loser is the winner, the man who, like my late, dear father, never has an accident, never is hit, just sits and swells in his driver's seat, from Willys-Overland to LaSalle, over the years his engines growing more and more powerful as he grows weaker and weaker. It was the second infarctus that got my dad. After the first, he watched the cholesterol, he stopped smoking, he thinned down until the confidence came back, he started driving again, eating, taking a taxi four blocks to the office. He thought he was whole again . . .

It is clear why Behrman calls his book 'Memoirs of an Autophobe'.

On reflection, it may be that our cars have become too dangerous for us as well, and not just for our children. Books like *Instead of Cars* offer us some hope that in future town planners may look more kindly on the bicycle. But then again we may have to wait until the last drop of oil has flowed before it comes to pass.

The point is, the bicycle should not be a toy or the dream machine of escapism. In this context, however, it is interesting to notice the

Summertime in England and the road to the coast.

Traffic duty in Peking, outside the Great Hall of the People.

fad for nostalgia in recent films and the part played in them by bicycles. Paul Newman performs pleasant antics in *Butch Cassidy and the Sundance Kid*. But the bike is only to inject a little novelty into cowboy scenery, and it becomes a prop to the film's calculated nostalgia for the old innocent days when even the crooks were somehow handsome fellows. In Truffaut's *Jules and Jim* the three lovers ride ecstatically round the south of France locked in a period of lyrical innocence and delayed adolescence – until, that is, their relationships begin to slip awry. Shades of Hoopdriver? Jerome K. Jerome knew the seduction of such feelings and he also knew their commercial power. The image-makers have always understood that our dreams of escape can be profitable in more senses than one.

'What bicycle did you say this was of yours?' asked George.

Harris told him. I forget of what particular manufacture it happened to be; it is immaterial.

'Are you sure?' persisted George.

'Of course I am sure,' answered Harris. 'Why, what's the matter with it?'

'Well, it doesn't come up to the poster,' said George, 'that's all.'

'What poster?' asked Harris.

'The poster advertising this particular brand of cycle,' explained George. 'I was looking at one on a hoarding in Sloane Street only a day or two before we started. A man was riding this make of machine, a man with a banner in his hand: he wasn't doing any work, that was clear as daylight; he was just sitting on the thing and drinking in the air. The cycle was going of its own accord, and going well. This thing of yours leaves all the work to me. It is a lazy brute of a machine; if you don't shove, it simply does nothing. I should complain about it, if I were you.'

When one comes to think of it, few bicycles do realize the poster. On only one poster that I can recollect have I seen the rider represented as doing any work. But this man was being pursued by a bull. In ordinary cases the object of the artist is to convince the hesitating neophyte that the sport of bicycling consists in sitting on a luxurious saddle, and being moved rapidly in the direction you wish to go by unseen heavenly powers.

Generally speaking, the rider is a lady, and then one feels that, for perfect bodily rest combined with entire freedom from mental anxiety, slumber upon a water-bed cannot compare with bicycle-riding upon a hilly road. No fairy travelling on a summer cloud could take things more easily than does the bicycle girl, according to the poster. Her costume for cycling in hot weather is ideal. Old-fashioned land-ladies might refuse her lunch, it is true, and a narrowminded police force might desire to secure her, and wrap her in a rug preliminary to summonsing her. But she heeds not. Uphill and downhill, through traffic that might tax the ingenuity of a cat; over road surfaces calculated to break the average steam-roller she passes, a vision of idle loveliness; her fair hair streaming to the wind, her sylph-like form poised airily, one foot upon the saddle, the other resting lightly upon the lamp. Sometimes she condescends to sit down on the saddle; then she puts her feet on the rests, lights a cigarette, and waves above her head a Chinese lantern.

Less often, it is a mere male thing that rides the machine. He is not so accomplished an acrobat as is the lady; but simple tricks, such as standing on the saddle and waving flags, drinking beer or beef-tea while riding, he can and does perform. Something, one supposes, he must do to occupy his mind; sitting still hour after hour on this machine, having no work to do, nothing to think about, must pall upon any man of active temperament. Thus it is that we see him rising on his pedals as he nears the top of some high hill to apostrophize the sun, or address poetry to the surrounding scenery . . .

... Down shady lanes, through busy towns on market days, merrily roll the wheels on the 'Bermondsey Company's Bottom Bracket Britain's Best', or of the 'Camberwell Company's Jointless Eureka'. They need no pedalling; they require no guiding. Give them their heads, and tell them what time you want to get home, and that is all they ask. While Edwin leans from his saddle to whisper the dear old nothings in Angelina's ear, while Angelina's face, to hide its blushes, is turned towards the horizon at the back, the magic bicycles pursue their even course.

And the sun is always shining. And the roads are always dry. No stern parent rides behind, no interfering aunt beside, no demon small boy brother is peeping round the corner, there never comes a skid. Ah me! Why were there no 'Britain's Best' nor 'Camberwell Eurekas' to be hired when *we* were young! ...

... You tired young man, sitting dejectedly on milestones, too spent to heed the steady rain that soaks you through; you weary maidens, with the straight damp hair, anxious about the time, longing to swear, not knowing how; you stout bald men, vanishing visibly as you pant and grunt along the endless road; you purple, dejected matrons, plying with pain the slow unwilling wheel; why did you not see to it that you bought a 'Britain's Best' or a 'Camberwell Eureka'? Why are these bicycles of inferior make so prevalent throughout the land?

Anyone who takes up cycling again, having last cycled as a child, will not find it effortlessly lyrical. Nor, with luck, will it seem nothing but misery and toil. It can be an odd and revealing experience, especially in contrast to travel by car. You may feel exposed and vulnerable to begin with, your hand twitching in search of a non-existent safety belt. The sense of exposure is not simply connected with the fact that cycling *feels* more perilous than motoring, it *is* pretty perilous. Yet it compares more than favourably with the motorcycle as far as safety is concerned. The Consumers' Association calculated that an ordinary person increased his death risk by about 10 per cent either bicycling or driving a car, compared with a 90 per cent increase for motorcyclists. (Looking at it another way, the cyclist could expect 80 years before an accident, the driver 125, but the motorcyclist only 12 years.) The sense of exposure involves a complete change in perspective.

The aim of the modern car manufacturer is to provide his client with a home from home, a tiny comfortable room with armchairs, thick carpets and, if he's rich enough, a stereo radio and tape

player. Have you noticed how secure and how isolated drivers feel? They do things in the middle of traffic jams – picking their noses or cleaning their ears – that they would scrupulously avoid in any public place. (Their cars are not called 'private saloons' for nothing.) If the car is expensive this comfort may be real; but only the very cheapest cars, and these usually French, do not aspire towards creating the illusion. In Britain the meanest Ford that goes is still a 'de-luxe'.

Robert Pirsig, in *Zen and the Art of Motorcycle Maintenance*, writes well of the difference between driving a car and riding a motorcycle:

You see things vacationing on a motorcycle in a way that is completely different from any other. In a car you're always in a compartment, and because you're used to it you don't realise that through that car window everything you see is more TV. You're a passive observer and it is all moving by you boringly in a frame. On a cycle the frame is gone. You're completely in contact with it all. You're *in* the scene, not just watching it anymore, and the sense of presence is overwhelming. That concrete whizzing by five inches below your foot is the real thing, the same stuff you walk on, it's right there, so blurred you can't focus on it yet you can put your foot down and touch it anytime, and the whole thing, the whole experience, is never removed from immediate consciousness.

This sense of immediate presence, of being part of the scene rather than a secure and private observer, is even more true for the non-motorized cyclist. It accounts also for the feelings of exposure and embarrassment experienced by the car-driver who starts to pedal. He feels self-conscious, the centre of the scene, stuck up on top of his machine, and therefore mildly foolish. When he climbs back down into his car he suddenly notices cyclists everywhere; they were always there, of course, but not the focus of public attention that he felt himself to be. Insulated from the road and the scenery as he is, the driver misses all the variety of his journey, the changes of temperature, the gradient, the smells and the road surfaces, and above all, paradoxically, speed. A cycle trip may incorporate forty miles an hour down a steep hill on a rough road and this will feel nothing at all like dawdling along a pleasant lane or crawling slowly to the top of another long incline. These sensations make speed real in a way that only occurs in a modern car at over eighty miles an hour. They may not always be pleasures exactly, but nonetheless they give cycling an intensity and an immediacy often lost to the driver.

The benefit which the cyclist gains in health and strength is enormous. The ergometer, a static cycling machine, is widely used in medical research as an indicator of physical capacity and efficiency because it exercises the heart, the lungs and the major muscles of the thighs and the back. As a cyclist you can expect to find that the hill that troubled you on your way to work bothers you less and less as you climb it each day. Cyclists call it 'putting miles in your legs', and it is extremely satisfying to experience an increasingly measurable and constant improvement in your performance, no matter how slowly you began.

To take such physical effort to its furthest extreme, professional cycle racing is the most demanding endurance sport in the world. This is especially true of the major stage races on the Continent – the Tours that last days or weeks and cover many hundreds of miles, including mountain passes, at an average speed of 20 mph.

The leaders of this sport are very rich young men. The Belgian rider Eddy Merckx, the greatest of them all, has a yearly income of about £217,000 and gets nearly £1,400 for entering a race. These sportsmen have something more than the charisma which surrounds professional footballers, because commercial sponsorship encourages the 'star system' and focuses a great deal of attention on the big names. This does not make their feats of tactical judgement and physical endurance any less extraordinary. One of the most outstanding and best loved cycle racers is the Frenchman Raymond Poulidor. His is a brilliant career spanning more than sixteen years with many notable victories. Yet he has never won the Tour de France, nor ever been race-leader in it, despite several stage wins and very good performances over the years. Yet 'the eternal second' is far from a loser and in 1976, at the age of forty, he finished an astonishing third in the Tour with a performance many younger men might well covet. When he won the Paris–Nice road race in 1972 – his thirteenth ride in this one-day classic – the newspaper *France-Soir* carried the news on equal terms with a surprise referendum just announced by the President. The front page was divided completely in half by two banner headlines 'POMPIDOU/POULIDOR', and the big photograph was of

Paris–Nice 1972. Poulidor dons the winner's jersey.

Raymond. One feels that 'Poupou for President' is not entirely a joke in France. Thus it is that a portly Continental businessman will still examine the equipment of a racing cycle parked in the street with much the same feeling that his British counterpart would reserve for a glimpse of a racing car on its trailer.

Let us look at the aesthetics of the machine itself. The bicycle has its proportions and sizes based on the human figure. It can be easily lifted, as befits its role as a helpful extension of human capacities. It is not a machine more powerful than its operator, nor is it a capricious and possibly dangerous animal. As an object the true bicycle is superbly functional. It is mechanically naked; nearly all its parts are clearly visible and easily comprehended. Lightness has always been at a premium in its construction because it uses only human strength to propel it. Therefore lavish styling or extravagant accessories have not interfered with its proper functioning. In this respect it has remained pure. Of course, functionalism is not the only aesthetic criterion, yet the machine which conceals inadequate, ill-designed or poor quality mechanics behind a smooth and appealing skin is a pretty inferior object.

A perfect example of the dangers of commercial 'styling' as opposed to engineering necessity can be found in the chopper-type bicycle or 'high-riser', that freak of modern design and selling methods which established high fashion in two-wheeled transport for ten-year-old pavement easy riders. This heavy, technically regressive small-wheeled machine resembles a parody of a motorcycle. Its gearchange, which need be no larger than a matchbox, has metamorphosed into a car-sized lever complete with shiny plastic knob. Indeed, the 'high-riser' is a toy carefully marketed to appeal as an ersatz motorized vehicle, a stopgap which pretends to provide some of the paraphernalia of speed before the child can graduate to the 'real thing'. Some of them seem deliberately balanced so that the front wheel can be lifted in the air like a dragster. As a monocycle this kind of bike provides the kids with thrills (and bruises), but it might have been designed to put them off cycling as serious transport and it certainly does not offer them the real physical demands and real rewards of cycling as a sport.

The true bicycle is amazingly efficient as a tool or machine. A man on a bicycle has the highest efficiency rating among all moving animals and machines. In terms of energy consumed in moving a certain distance as a function of body weight it is far more efficient than a man walking, and it outstrips jet planes, horses and even the sleek salmon.

Put it another way: if the energy a cyclist needs from food is compared with the energy from a gallon of petrol, the cyclist is capable of 1,600 miles per gallon. The bicycle utilizes the most powerful muscles of the body in the best possible motion, a smooth rotary action. A human being consumes energy simply by standing up, and even walking uses muscles that do not actually aid the movement of the body along the ground. Thus, for example, the leg muscles must not only support the rest of the body but also raise and lower it continually, consuming energy but not performing any externally useful work. The cyclist is seated, however, and this relieves the leg muscles of their supportive function. The power of these muscles is used as efficiently as possible because the rising leg is lifted by the downward thrust of its fellow. All that must be overcome is wind resistance and the force of gravity on hills, and this is what drop handlebars and gears are for. This leaves the rolling mass of the machine itself, and with high pressure tyres on a smooth surface this is very little. Thus on a modern lightweight a man's energy consumption for a given distance is reduced to about a fifth (approximately 0.15 instead of 0.75 calories per gram per kilometre), while at the same time he is able to travel three or four times faster than walking speed.

The bicycle is kind to our dwindling world resources (according to the Consumers' Association, the energy needed to make one car would make a thousand bicycles), so it is not surprising to discover that Ivan Illich applauds it in the essay which forms part of his 'epilogue to the industrial age', *Tools for Conviviality*.

> Tools foster conviviality to the extent to which they can be easily used, by anybody, as often or as seldom as desired for the accomplishment of a purpose chosen by the user. The use of such tools by one person does not restrain another from using them equally. They do not require previous certification of the user. Their existence does not impose any obligation to use them. They allow the user to express his meaning in action.

By these terms the car creates systems which menace as many or more people than they serve and thus it is anticonvivial. Congestion in our cities, concern with pollution, the world-wide energy crisis and the rising cost of oil have all served to make us think twice about the motorcar and reappraise the bicycle, at least for town transport. Pedal power also lends itself quite naturally to the need felt by many people for proper diet and healthy exercise.

In recent years there has been a considerable revival of interest in cycling in America. Riding a bike is potentially much more dangerous in the States than in Britain, nevertheless it has attracted the enthusiasm of a wide variety of people and New York businessmen pedal happily alongside Friends of the Earth. The machine they both aspire towards may well be a hand-built English lightweight. The Americans take their cycling seriously and the literature is full of informed and weighty articles about training, diet, carbohydrate loading, the strength of different alloy frames, the psychology of long-distance touring, and so on. Are these footnotes to a passing fad or genuine indications of a lasting change in transport priorities? The question needs careful consideration.

'. . . symbols for the keen ecologist'.

In Britain fashion is not far behind. In 1976 a long-established Austrian cycle firm introduced a top market superbike to the bemused columnists of the London news media:

> Top of the tree, the £599.95-and-keep-the-change-my-man Ultima Superleicht is the newest status symbol for the keen ecologist. Cheaper to run than a Rolls-Royce but just as exclusive, the Ultima is guaranteed to make heads turn wherever it is seen.

It is a beautiful bike, though many clubmen own machines that are just as refined yet far less extravagantly expensive. The 'pampered executive' for whom it was intended might not have found its outright racing lines particularly comfortable, but at least for the price he was provided with a foam-filled case for garage-wall mounting, perhaps to obviate the necessity of ever riding his purchase. It seems that 'ultimates' have never had much to do with practicality.

If cycling has become fashionable in Britain and America, then the Hoopdrivers may have the laugh on Joe Lampton after all. Trained by those years of practical austerity transport, Hoopdriver's legs will propel him effortlessly, even 'lyrically', up the steepest hill. But trendy Lampton may find getting to the top this way a bit harder.

2 The Bicycle Today: Naming the Parts

Most people think that there is not much to choose between one bike and another. They have a Platonic Idea of two wheels, a saddle to sit on, handlebars to hold and pedals to make it go. They are right in principle, of course, but the details may well make them wrong in practice. Most bikes readily available in the shops are perfectly good, solid, functional machines. But for every purpose there is a best bicycle, and a machine which is wrong for your needs will simply make cycling seem a chore.

How to begin? How should one choose between the innumerable variations of the basic bicycle idea? Why are some bicycles more than ten times as expensive as others? What are the features to look for and understand?

Unfortunately (and as usual), knowledge comes only with experience. It is only when you have finished making a bookcase or mending the plumbing that you know how you should have set about it in the first place. So it is not until you have bought a bike, used it in different conditions and considered the whole business in the past that you will really know for sure what you wanted. Still, bicycles are quite easily borrowed for test-rides, they do not need insurance or driving licences, and with reasonable care they are all but idiot-proof. Nothing can match practical experience and experiment, but this chapter will provide a starting-point, mainly by straightforward description. In the army it is called Naming the Parts.

THE WHEELS

First of all, the wheels. In the early 1960s an independent engineer and inventor, Alex Moulton, caused a stir by designing and manufacturing a small-wheel bicycle. Since that miniature revolution the public has had to choose between two apparently widely differing shapes of bicycle. Fashion has more or less favoured the small-

Alex Moulton launching his mini-bike in 1963.

wheeler; tradition in the cycling fraternity sticks with the old design. The debate is interesting and worth exploring because it involves certain basics in bicycle lore and technology.

The main advantage of the small-wheeled cycle is its low centre of gravity. The heavy members of the open frame are all near the ground. So far so good. This makes such a machine stable, apparently more easily manoeuvrable, and especially good for carrying loads. A top-heavy rider is a dangerous proposition and folk who cycle with rucksacks perched high on their backs, for example, are inviting trouble in the form of wobbles and swerves. However, small wheels have one easily understood disadvantage. They cannot smooth out road surfaces with quite the same effectiveness as their more ample cousins. Consider the logical exaggeration of the small wheel idea: a bicycle with tiny wheels. Roller skates are no good at all except for the smoothest surfaces. A 27-inch wheel, the normal size, will roll easily over a rut which would snag a roller skate utterly. The diameter of a wheel is an important factor in shock-absorption. In fact, the Ordinary or penny-farthing was in this respect a design surprisingly well-suited to roads rutted with pot-holes and hazardous with stones. Its huge, slender wheel, 50 inches or so in diameter, had a habit of dumping you on your head when going downhill, but that is a different matter. There are other factors which govern the handling of bicycles on rough roads, in particular the design of the frame and the length of the wheelbase, which are discussed elsewhere. However, one further aspect of the small wheel which makes it bad on bumps is its rigidity; short spokes make stiff and unforgiving wheels.

Mr Moulton solved the problem of shock-absorption in his new design by incorporating sprung rear and front forks in his frame, an exciting and intriguing new development. The next question, however, is why complicate the rigid frame of the bicycle with squashy rubber cushions, springs and hinges, all of which introduce new points for wear and tear and new weaknesses? Raleigh asked this question when they took over Moulton in 1967, and phased out the original suspension. By 1975 the only cycles with the Moulton name were produced for children, and the original design was already becoming a collectors' item. Other small-wheelers now solve the bumpy ride problem by utilizing fat squashy tyres to sort out uneven road surfaces. But again this contradicts a fundamental aspect of cycle design. The more tyre-surface on the road, the more friction, and the more work needed to turn the wheels. Riding a small-wheeler with bungey cushion tyres after being used to a more sporting wheel and tyre feels something like pedalling through glue. Many manufacturers have therefore compromised on the small wheel with a medium-sized 20- rather than 14-inch diameter, and this seems a sensible solution, preserving some of the benefits of the original idea, such as the good luggage-carrying capacity, but not involving a departure from the rigid frame.

The Moulton supplied a vital surge of new interest in the bicycle just at a time when business was stagnant and ideas were stale. No one worried that the small wheels, revolving nearly twice as fast as normal, also meant more tyre wear and more wear of the hub-bearings. It looked new and neat; it matched the mini-car and the mini-skirt in a Britain concentrating on the virtues of compactness and turning against grandiosity. It had the right spark of style for its time and, even better, it set the industry and the enthusiasts thinking and arguing. Folding bicycles, which had been on the fringes of the cycle world since the First World War, became a practical reality for the first time. These machines could be kept in a small flat, or stored in a cupboard or car boot, where they might moulder peacefully until their owners gathered the energy to fiddle them back into rideable shape. Unfortunately, their natty appearance makes them no easier to carry up stairs than the normal bike, and, indeed, the need for hinges and clamps makes them heavier than their image suggests. There is an ultra-light small-wheel folder on the market, the Bickerton, but its all-alloy materials make it rather expensive. It is slightly less rigid than a steel machine, but it is a serious attempt to solve the problem and British Rail apparently accept it as 'hand luggage'.

In the end, however, the virtues of the small-wheeler lie chiefly in its fashionable appeal and its usefulness for shopping. Like scooters in relation to motorcycles (remember the Vespa?), they look middle class somehow, respectable and different. They are perfect for small round-town trips, but are hardly ever seen beyond the suburbs. Some Moultons were successfully converted for racing and touring because their suspension system allowed the use of lightweight high-pressure tyres.

Lightness is obviously of major importance in cycle construction, leading to basic distinctions between different kinds of bicycle. The

The Bickerton has steel handlebars and chainstays, but extensive use of alloy everywhere else keeps its weight below 25 pounds. It is genuinely portable and with practice it can be unfolded in about 30 seconds.

most important factor is revolving weight, that is to say, the weight of the moving parts. A ten-pound weight in the saddlebag will add to the bike's dead weight when climbing a local Alp or pushing the bike up to cruising speed. Imagine the same weight distributed around the bicycle, attached to the pedals, or, most important, to the wheel rims. The result would be a machine needing much more effort simply to keep these parts in motion, far less propel the bicycle forwards. If you are two stone overweight you may wobble a bit when you run, but if that extra weight was all in your feet, fourteen pounds in each shoe, you would hardly be able to run at all. Consider also that, as the diameter of the wheel rim is larger than the circle described by the pedals, weight at the rim is acting as a lever against the force exerted by the feet; and this effect is magnified by the gearing of the bicycle, which will make the rear wheel revolve perhaps twice or three times to every revolution of the cranks. By comparison, the dead weight of the frame, saddle, mudguards and so on is supported restfully on the ground.

Revolving mass is therefore of prime importance, and the most vital factor in it is the weight of wheel rims and tyres. Consequently, in order to discriminate between bicycles, the first step is to be able to recognize the different kinds of wheel and tyre used nowadays.

There are three kinds of wheel rim in common use. The heavy-duty, almost flat Westwood is steadily becoming a rarity. It is almost always found with the old-fashioned roller lever brakes, which have an arrangement of rods, hinges and levers running around the bicycle like some primitive central-heating system. There are advantages to this method, however. The rods, unlike cables, are virtually impervious to wear and are likely to last a lifetime of constant use. The brake blocks are pulled upwards into the inside of the wheel rims, rather than closing on the sides of the rims, and this provides just as good a braking force as the more modern cable brakes. Westwood rims are not light, however, and they support a heavy $1\frac{1}{2}$-inch-wide tyre, plus inner tube. This was a virtue on Edwardian roads and it still is for parts of the world where the going is bumpy, but it all adds up to extra load for riding on tarmac.

The most common rim, the Endrick, is also used with a standard 'wired-on' (or 'clincher') outer tyre, plus inner tube. These can differ in width ($1\frac{3}{8}$ inches or $1\frac{1}{2}$ inches) and marginally in depth, and although these differences may seem tiny, wheels of this sort vary enormously in weight and riding quality. The narrower rims are designed for use with high-pressure tyres, which should be pumped up to around 60–90 pounds per square inch instead of the normal 40, thereby decreasing the friction with the road surface and creating faster, lighter and more responsive wheels. Even better, the narrow Endrick rim is available in alloy rather than steel and chrome, not only making the wheel much lighter where it really matters but also increasing braking efficiency: alloy rims combined with soft rubber brake blocks are much better for stopping in the

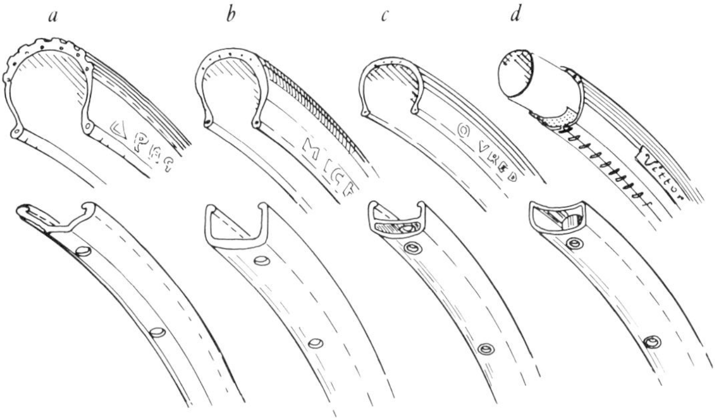

Fig. 1. Wheelrims and tyres.
a. Westwood rim and roadster tyre. *b*. Endrick rim and medium-weight HP tyre. *c*. Super-narrow rim and HP lightweight tyre. *d*. Sprint rim with racing tubular. Note that the inner tube is completely encased inside the sewn-up cover.

wet, since water on a chrome surface seems to act as a natural lubricant for hard rubber blocks. These alloy rims, though more expensive than steel, are a very worthwhile feature to look out for on a bike – they do not rust, either.

Some companies now make a very narrow high-pressure tyre for the enthusiast, and there are special 19-millimetre rims to fit them. These light tyres should be inflated to pressures around 90 pounds per square inch, and their handling qualities are very like those of the racing man's tubular. They are ideal for the rider who demands a totally responsive machine at all times. We would recommend them for the beginning racer, too, for they are cheaper and easier to repair than tubulars. They need to be treated with respect, however, and a heavy skid will rip through them.

Finally, the sprint rim with the tubular tyre. Instead of being kept in place by wires on the outside edges of the outer cover, these ultra-lightweight tyres have to be cemented with special glue to a shallow, narrow (about $\frac{3}{4}$-inch), concave rim, which is always alloy. Some rims intended for track racing have a wooden internal section for featherweight wheels. Tubulars, or 'sew-ups', or 'singles', as they are sometimes called, are pumped up to pressures of between 90 and 120 pounds per square inch, thus giving a small contact area and minimal friction with the road. They can be very expensive; they are impossible to mend on the spot and comparatively difficult to mend at all, so a spare tyre has to be carried always. Many experienced cyclists use them for touring and general-purpose riding, but they are probably best left to the confirmed enthusiast or would-be expert. Certainly it would be unwise to buy a bicycle ready fitted with tubulars unless you were sure that your local shop sold spare 'tubs'. The riding characteristics of sprint wheels are marvellous and they are essential for racing. For most general purposes, however, their advantages are outweighed by their disadvantages.

Between rim and hub are the spokes. You don't have much choice here, unless you have wheels specially made. British bikes used to have 40 spokes in the rear wheel and 32 in the front. Almost all ready-made bikes now use the 36/36 Continental system, with either chrome or rustless steel spokes. (If you have a choice, 'rustless' are supposed to be less brittle than chrome spokes.) Racing cyclists build wheels with only 28 or even 24 spokes, often double-butted, which means that they are reinforced by a thicker section at the hub and the rim. Such super-light wheels are unnecessary except for the special purpose of racing, and a pair of the best, with tyres, can cost more than a whole bicycle of the standard variety. They are the most important item on a racing machine – its heart, if you like.

Hubs can have small or large flanges. The large flange, as it leads to shorter spokes, creates a stronger wheel and is almost standard equipment on the better-quality lightweight cycles. Racing cyclists on the long stage races favour the small flange for its increased comfort.

Quick-release hubs, a racing refinement finding its way on to the general-purpose market, are useful devices which allow a wheel to be removed in seconds by simply turning a small lever through 180 degrees. Accurate replacement of the wheels in the forks is also much simpler and faster. Wing nuts also simplify this particular feat, and you don't have to fiddle around with a pair of spanners. It is pleasant to be able to zip off a wheel and pop your bike into the back of a car. You should check these Q-R levers or the wing nuts from time to time, however, to ensure that the wheel is still firmly in place: they are far from fiddle-proof. They are also, of course, a thief's delight.

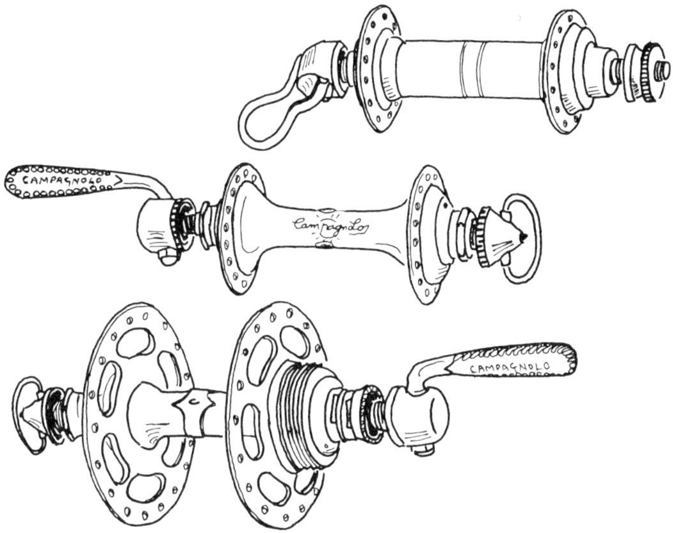

Fig. 2. Three quick-release hubs: a sealed-bearing front hub, a small-flange front hub, and a large-flange rear hub.

All parts of a hub – the spindle, cones and ball bearings – can readily be replaced when worn. There is a current trend, apparently started in America, for 'maintenance-free' (or irreparable) sealed-in hub bearings. The best of these are very expensive, and ought technically and morally to last for the life of the bicycle. Cheaper versions have not been on the market long enough for anyone to know about their durability. It is a moot question of cycling ethics whether bicycles should be divisible into basic components which are readily available and easily renewable, or whether whole units should be unmendable and have to be replaced in their entirety by a dealer. Ordinary hubs are by no means foolproof; they are often badly adjusted, so that riding a new bike without checking the hubs can ruin the bearings in a few minutes. But it seems reasonable to deplore the tendency for cycle parts to come in larger and more expensive chunks. This 'replace the unit' philosophy saves money for the workshop owner (in mechanics' time and wages) but not for the home repairer.

THE FRAME

If the wheels are its heart, the bike's skeleton is its frame. Small-wheel cycles generally plump for a more or less open frame. As there is only one tube between saddle and handlebars it has to be strong, and to be strong it has to be thick and heavy. The conventional frame on the other hand is an elegant, airy, rigid and robust way of uniting the extremities of the cycle. It is by no means a haphazard design, for it has been gradually improved and changed to suit new road conditions over the last three quarters of a century. And even within this familiar layout there are subtle differences and refinements which are worth knowing about.

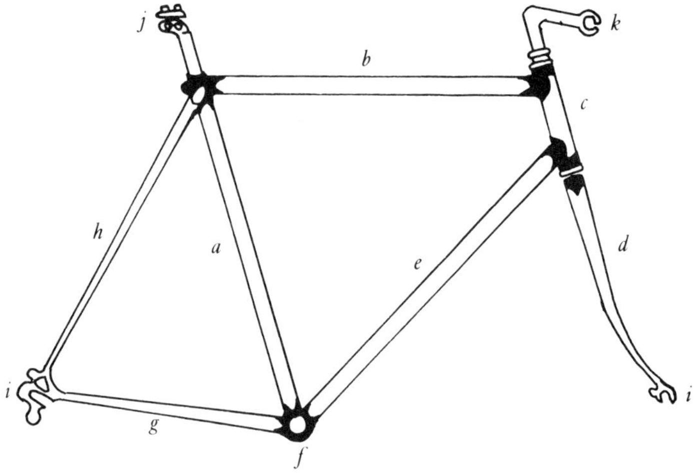

Fig. 3. The modern diamond frame and its parts.
a. Seat tube (the length of this gives the size of the frame). *b*. Top tube. *c*. Head tube. *d*. Front forks. *e*. Down tube. *f*. Bottom bracket. *g*. Chainstays. *h*. Seat stays. *i*. Front and rear drop-outs. *j*. Seat pillar. *k*. Stem.
(Black areas indicate the lugs, the fork crown and the bottom-bracket shell.)

The most common kind of frame is made from lightweight steel tubing, joined at the relevant corners by being brazed into clusters of shaped and pressed steel called lugs. Some engineers believe that lugs are a weight-expensive answer to an originally bad design and that the triangles inherent in the frame's shape do not in fact strengthen it at points which bear the maximum load. However, the diamond frame has stood the test of time, and none of the variations drawn up by theorists have gained much general support. In some frames the tubes are 'welded' together without the lugs. This method of construction is to be found among both the cheapest and the most expensive cycles, but not much in the middle range. Almost all cycle 'welding' is, in fact, brazing. Cheap joints are made by tacking tubes together with a blob of molten brazing compound. The best work involves building up a seamless joint between two mitred tubes and then rubbing down the finished work until it is satin smooth. This method is used in some time-trial bikes to save weight.

'Ladies frames' are available, built without a horizontal top tube so that they can be straddled while wearing a skirt, but this design sacrifices the rigidity of the diamond frame and it is never used for racing. The 'mixte' frame is said to be more rigid than the more common twin-down-tube style, but nowadays many women prefer to ride the so-called 'gent's frame'.

If you are much smaller than average you may have difficulty in finding a diamond frame to fit you. Few manufacturers go below 20½ inches frame size in adult bikes, and even then these are often scaled-down versions of the larger models, so that their angles and proportions are all wrong. If they are enthusiastic riders smaller women and men will order hand-built frames to fit them, and a steeper seat tube and a shorter top tube will ensure that they are not stretched out along the machine. But this costs more and many cyclists regret that a properly proportioned small diamond frame is not available 'off the peg' from the major manufacturers.

The best kind of tubing for framemaking is cold-drawn seamless, either plain gauge or double-butted. In a double-butted tube the walls are thinner in the centre section but thicker and stronger inside the ends around the point of entry into the lug, the place where breakage is most likely to take place. The most celebrated make of this kind of tubing is the product of the British firm, Reynolds, and is called 531 Double Butted tubing. Cycles made of these tubes will carry a small label stating whether only the main tubes or all of them, including forks and chainstays, are Reynolds 531 D.B. or 531 Plain Gauge. Other acclaimed makes are the French Vitus or Italian Columbus tubing. Some ready-made bikes are made with slightly less refined plain-gauge tubing, but are nevertheless perfectly strong and serviceable. The advantage of lightweight double-butted tubing is that it achieves a perfect combination of rigidity with shock-absorbing qualities. You also have the satisfaction of knowing that you have the best, along with a few ounces off the overall weight of the bike. Most off-the-peg machines are made of seamed tubes, in which the steel has been rolled into a tube and welded (invisibly) along its length. If you put your finger inside one of these tubes you can feel the join.

Minute changes in the angles and dimensions of the frame give the bicycle remarkably different handling qualities, and this frame geometry is explained in Chapter 3. Most mass-produced cycles have a fork rake of about 2½ inches and frame angles of 72 degrees. Variations in frame geometry are subject to fashion and taste, but it is worth knowing why a friend's bike might feel so completely different from your own when they look identical to the eye. And if you intend having a frame hand-made to your own requirements it is vital to give the matter some consideration.

Small stops, eyes and tunnels can be brazed on to the frame for the brake and gear cables to pass through, eliminating their tendency to bend and flex in use. This keeps the lever controls positive and taut, which is especially important if spongy braking and gear changes are to be avoided. Fashion is at present against too many brazings on the tubes, as the heat necessary to apply the brazing is supposed to weaken them, even though low-temperature techniques tend to make this objection theoretical. The current trend is an Italian preference for cable clips, even although these are liable to scratch the paint or slip. Of course brazings can break off too, but they should not.

SEAT PIN AND HANDLEBAR STEM

Two items which might almost be considered part of the bicycle's frame are the stems to which the saddle and handlebars are attached. Saddle pins, pillars or pegs are manufactured in steel or alloy; heavy riders might opt for steel, as a broken saddle pin could obviously be very dangerous. The small-wheel cycles, with their open frames, often employ extremely lengthy pillars which have to bear the whole weight of the rider without the support of top tube or stays. This is an undesirable feature of their design and one which can be dangerous, for a long-legged rider may set the saddle so high that the stem will be insufficiently contained by the seat tube. Many small-wheelers have a limit mark clearly indicated to avoid this situation. For safety the pin should always have at least three inches inside the seat tube. Usually the saddle is attached to its pin by an awkward and crude little clip affair, but the very best seat pillars incorporate a micro-adjustable cradle.

Handlebar stems or extensions come in a variety of shapes and sizes and can also be made of steel or alloy. Sprinters, who pull heavily on the handlebars in order to accelerate fiercely, sometimes use a steel stem, but for ordinary use alloy is quite strong enough and much lighter. However, you should not compromise on quality, as any break would lead to a bad accident.

CHAINSET AND GEARS

After wheels and frame, the bike's transmission is the next most important item for consideration, and here there is a wide variety of choice as this includes the cycle's gears as well. The transmission involves cranks, chainwheel (in combination called the chainset), chain, the rear sprocket and freewheel mechanism, gears and, ultimately, the pedals.

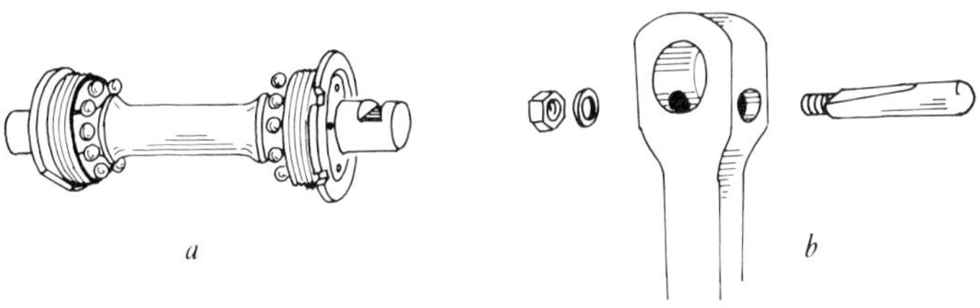

Fig. 4. A bottom-bracket set for cottered cranks.

a. The ball bearings run between the inside of the cups and the flared bearing surface on the axle. The cups thread into the bottom-bracket shell on the frame and one of them can be adjusted and secured with its locking ring.

b. Cotter pin and crank. The flat on the cotter pin engages with the groove cut across the end of the spindle and the pin is then driven home like a wedge.

Running through the bottom bracket shell, and anchored in place by cups and ball bearings allowing it to turn freely, is a spindle of forged steel to which the cranks are attached. When the rider stands on the pedals, the forces acting on the spindle are enormous, the weight and the strength of the rider being multiplied by the leverage of the cranks. It is thus vitally important that the cranks and the axle should be firmly attached to each other, and the usual method of achieving this is crude but moderately effective. The primitive and temperamental cotter, a tapered pin with a flat edge, is slipped through holes in the crank and rammed home against flat sections on the axle, to be held in place by a nut on the other side. If in the process of time and use this pin works loose, then an annoying click occurs at every turn of the pedals and energy is wasted. The fact that the crank can come loose on the axle does not mean that the cotter pin can easily be removed and they are sometimes diabolically difficult to displace. Somehow it always comes down to blind rage and a big hammer. (See the comments on maintenance at the end.) Expensive bicycles have cotterless cranksets. Each end of this type of spindle has a tapered square section. The crank has a matching square hole, and it is kept jammed tightly in place by a bolt screwed into the centre of the spindle. Such cranks and their chainwheels are made of alloy and so are strikingly lighter than their cottered steel cousins. And cranks, of course, are part of the bicycle's revolving weight.

The up-and-down motion of the feet on the pedals is converted into rotary motion by the crankset and transferred to a smaller sprocket on the rear hub by means of the chain. The ratio of size between the front large sprocket (the chainwheel) and the rear sprocket governs the number of times the wheel will turn for every revolution of the pedals. One way of expressing this ratio is to imagine that the effective diameter of the rear wheel has been increased. On the old large-wheeled Ordinary the size of the front wheel, and thus the speed of the bike, depended on the inside leg measurement of the rider, giving a wheel of 50 or even 60 inches in diameter. The early racers tended to be big men. The gearing of modern cycles in Britain and America is still expressed in terms of the imaginary effective diameter of the rear wheel. To calculate a gear size, therefore, the diameter of the rear wheel is multiplied by the number of teeth on the chainwheel, and divided by the number

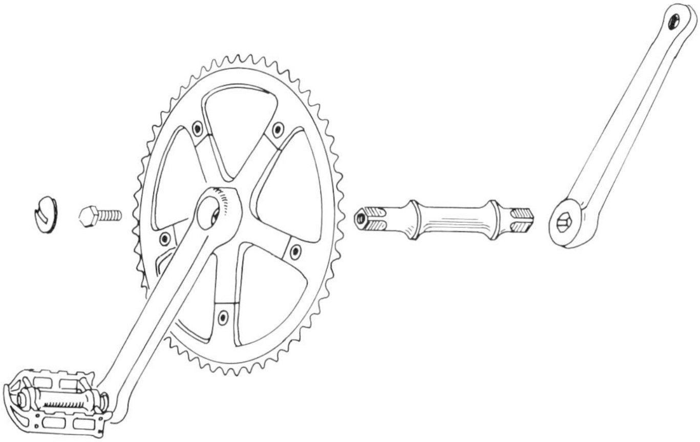

Fig. 5. A cotterless crankset. Each crank fits on to the tapered square end of the spindle and is held in place with a bolt. The cranks, the chain ring and the quill pedal also shown are all in alloy, but spindles, bolts and bearings are still in steel.

of teeth on the rear sprocket. For example, a bicycle with 27-inch wheels, and with a 45-tooth chainwheel combined with a 16-tooth sprocket, will have a gear of 75.9 inches:

$$\frac{45 \times 27}{16} = 75{\cdot}9$$

Gear charts will mean nothing to you unless you already have a bicycle and some experience. But, if you think of a 100-inch gear in terms of a back wheel over eight feet high to be turned once by each turn of your pedals, you will begin to guess how it feels. As usual, it is not until you have a bike that you know exactly what you require in the way of gears and by then it may be too late. But some information and forethought can still be useful.

First, however, how are the ratios varied on a bicycle in order to cope with different loads, gradients, wind resistance, and the varying strength of different riders? There are two kinds of gear. The hub gear is a complicated arrangements of cogs contained in a fat rear hub. The normal ratio, or direct drive, can be worked out by using the formula above. Variations on this are fixed percentages above or below the normal ratio. A Sturmey Archer hub gear with three speeds gives a low which is $\frac{3}{4}$ of normal drive and a high which is $1\frac{1}{3}$, while the rather rare five-speed version used to provide lower gears of $\frac{2}{3}$ and $\frac{15}{19}$ and higher gears of $\frac{19}{15}$ and $\frac{3}{2}$.

The other method of gearing, the derailleur, is more variable and can provide wider ratios than the hub gear. On the body of the freewheel is a cluster of different-sized sprockets, nowadays usually five. The chain passes through the rear changer, a mechanism which derails the chain by a side-to-side movement, allowing it to move from cog to cog while you are pedalling. The derailleur also keeps the chain tension even, so as to accommodate the different-sized sprockets, by means of a sprung jockey arm which can move backwards and forwards. The number of gears can be doubled or even tripled by using twin or triple chainwheels along with the multiple rear sprockets. The front changer shifts the chain across from chainwheel to chainwheel. As the rear changer keeps the chain

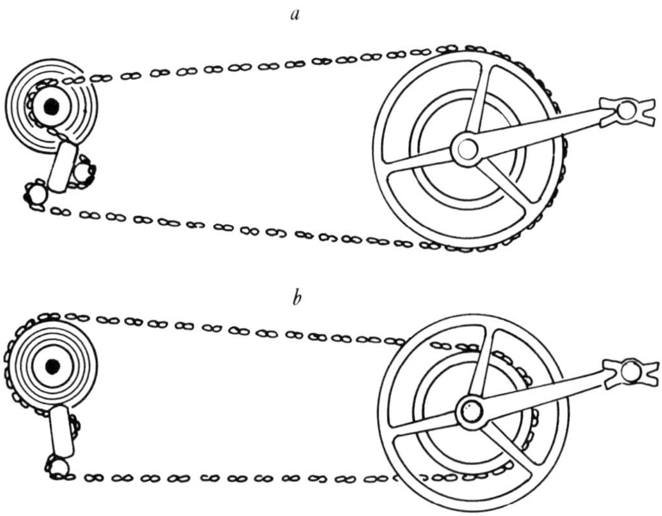

Fig. 6. Gear changes on a ten-speed derailleur system.

a. The highest gear is given with the chain on the big chainwheel and the smallest of the five rear sprockets. This is a gear for level roads with the wind behind you.

b. The lowest gear is used for climbing hills with the chain from the inside chainwheel to the largest rear sprocket (also on the inside). Extreme cross-overs (big to big, small to small) should be avoided.

taut it is not possible to have a double chainwheel (or 'double plateau sprocket', as the Americans call it) without a rear derailleur or some chain tensioner.

What are the relative advantages of the two types? The hub gear is protected from damp, dirt and knocks and is very reliable, though if it does go wrong it is a job for experts only. The gear change is achieved by manipulating a small trigger which can be placed under the rider's finger on the handlebars, or even incorporated into a handlebar twist-grip. Gear changes should be made while *not* pedalling, a slight disadvantage when going uphill, but good at traffic lights or when starting off. The chief disadvantage is that hub gears provide neither very wide ratios from high to low nor a pleasant progression of ratios in easy steps. An average set-up might be: a 46-tooth chainwheel combined with an 18-tooth sprocket and 26-inch wheels, giving gears of 49.8, 66.6 and 88.5 inches. This is perfectly good for riding around and even long-distance cycle tourists might find it adequate. Two facts do still count against it however: it offers no choice of ratios, and when in high or low the rider is having to turn a series of cogs inside the hub which create extra friction and more work. The derailleur uses an almost direct drive by comparison and you can fit practically any gearing you want. Nevertheless, the hub gear is sometimes unfairly maligned by club men, and for rough riding or town commuting, built into a lightweight wheel, it compares surprisingly well with the derailleur. The differences in weight and friction are not so great as some suppose. The now rare fixed-wheel hub gear is still treasured by a few *aficionados*.

The derailleur is characteristically the gear used by cycling enthusiasts. Part of its appeal may be that it is mechanically naked; it is easy to see how it works just by looking at it carefully and it can be taken apart and repaired with relative simplicity. The derailleurs are manipulated by levers usually situated on the down tube of the frame, but sometimes at the ends of the handlebars, on the top tube or near the handlebar stem. Safety experts have argued that leaning slightly forward to change gear is perilous, and that the finger-tip control trigger of the hub gear is preferable. Hub gears are certainly more suitable for children for this reason, and also because they are less vulnerable than the delicate jockey mechanism of the derailleur, which will not survive knocks or being dropped on the ground.

Furthermore, derailleurs can only be changed when the pedals are turning the sprockets, so if you stop at the traffic lights in high gear, getting going again can be difficult. You soon learn to think ahead and change down.

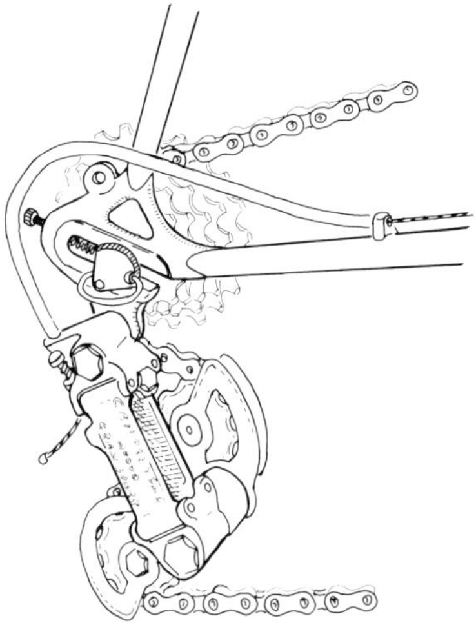

Fig. 7. Rear drop-out and typical derailleur. Forged drop-outs with rear adjusters are found only on quality frames. The round eye is threaded to take a bolt for the mudguard stays. Note the brazed-on cable stop.

What gear sizes suit what purposes? Long-distance tourists addicted to 'pass-storming' and carrying luggage will opt for a set-up which achieves very low gears, perhaps even less than 30 inches. Chainwheels of 30 and 46 teeth combined with a sprocket block of 14–17–20–24–28 teeth will provide a fairly evenly spaced set of gears with little duplication ranging from 28.9 to 87.7 inches. At the other extreme a racing cyclist planning for a road race will want large gears and a close-ratio sprocket block for positive and quick shifts: 42 and 52 teeth on the chainwheels plus a 13- to 18-tooth block will give gears from 63 to 108 inches. For average use a gear of 30 inches is too low and 110 inches far too high. It is very difficult to recommend gear ratios. Anyone who has ever tinkered around with different-sized sprockets or even contemplated the gear charts with this in view quickly realizes that there is no ideal set-up that will suit all circumstances. Around 40 inches will get up most hills with a push, while 100 inches can produce speeds of over 40 mph on moderate downhill stretches – quite fast enough for the average rider.

A word of warning: most production cycles with derailleurs are geared too high for the beginner, with chainwheels of around 48 teeth and sprockets only running up to 22 or 24 teeth. The dealer should be able to replace the block and provide a wider ratio quite easily. A bigger sprocket at the back will make those hills much easier. Also, ten-speed systems almost always involve some duplication of gears, and occasionally tedious double changes might be needed in order to run through all the gears. These drawbacks are avoided without much loss if you stick to five gears only. The better the equipment you buy the more potentially variable will be your transmission. Any special gear set-up, such as for Alpine touring, will almost certainly require better than average equipment.

The bicycle chain is a complicated though often neglected item. Derailleur chains need to be more flexible than the ordinary chain and are therefore thinner: $\frac{3}{32}$ inch wide, compared with the average $\frac{1}{8}$ inch. The wider chain has a spring link for easy removal, while the derailleur chain requires a special, inexpensive tool for unriveting the links.

Pedals are the final essential items in the transmission. These can be made of thick rubber, but this compresses in use, thereby wasting energy, and rubber pedals are heavy. The lighter rat-trap or quill pedals, made either of steel or alloy, are essential for anything but pottering. Rubber pedals can also be slippery when wet, but then leather-shod feet can slip on metal pedals too. Toe-clips and straps, which can only be used with metal pedals, are in a sense the final link in the transmission. They keep exactly the right part of the foot in contact with the pedal so as to maximize leg power, but, above all, they prevent feet slipping off pedals, which can result not just in a cracked shin but in complete loss of control of the bike.

Toe-clips look dangerous, and most non-cyclists dislike the idea of strapping their feet to the pedals. Except in racing, however, it is unnecessary to tie the strap tightly, and the clip can be flipped on or off as easily as a riding boot can be removed from a stirrup. In other words, there is a knack to it. It would of course be foolish to equip a child's garden racer with toe-clips, and in stop-start riding in town they can be a nuisance. On balance, however, they are more useful than not and, if anything, they are a safety feature. Practice makes their use effortless and speedy – if you are worried by them, use them for a while without the straps until you get the hang of it. For the serious rider they are essential.

THE BRAKES

Bicycle brakes are at present receiving a good deal of criticism for their inefficiency compared with those of cars and motorcycles. One problem is that super-efficient brakes would be just as likely to cause accidents as prevent them. Hub or coaster brakes are fitted to some bicycles in Britain and have always been common on the Continent; they can stop a rear wheel revolving very quickly, so quickly indeed that they are likely to make the bike skid on greasy road surfaces. They are in fact quite difficult to apply gently. The more usual kind of brake, consisting of claw-like calipers operating on the wheel rim, are fine in the dry but can be spongy and inefficient when steel rims and hard rubber brake blocks are covered with a film of water. Bikes are more manoeuvrable than cars, and they can escape danger even through alarmingly small gaps. But braking on a bicycle needs consistent care and constant forethought. Safety experiments to date have often led to the conclusion that braking effectiveness depends more on the skill of the rider than on equipment of different kinds or qualities. And at least the cyclist does have an independent braking system on each wheel. Cyclists can only hope that manufacturers will continue to experiment with brake block materials and different methods of braking in order to find the perfect means for slowing the bike down, evenly and powerfully, but without the snatch that leads to skidding.

For the prospective buyer there is not a wide choice. All the alloy caliper brakes are just about as efficient as each other, provided that they are well adjusted. As has been mentioned, alloy wheel rims provide a better braking surface than steel. Centre-pull rather than side-pull calipers are easier to adjust, and, as the arms pivot nearer the brake shoes, they provide greater leverage through the arms to which the brake shoes are attached.

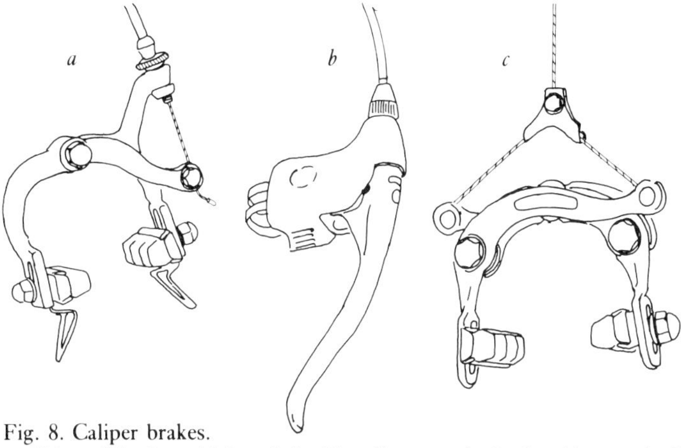

Fig. 8. Caliper brakes.
a. Side-pull brakes with knurled cable-adjuster and wheel-guides attached below the brake blocks. The calipers pivot on the central fixing bolt.
b. Brake lever for drop handlebars. This lever has a rubber hood for comfort and a knurled cable-adjuster.
c. Centre-pull brakes. The twin actuating arms have separate pivots lower than the central fixing bolt. The cable 'bridge' ensures an even pull on both brake blocks.

Tandems, on which the brakes need to stop twice the weight, sometimes use cantilevered brakes, providing even more leverage through the arms to which the brake blocks are attached. Unfortunately these require brazed-on attachments to the forks, and so they are only really available to those who have a bike hand-made to their own requirements.

The brake levers themselves are all equally efficient provided they are of the right kind for the style of handlebars to which they are fitted. Switching from dropped to straight handlebars should not be done without changing the brake levers too.

THE HANDLEBARS

The handlebars of a bike are the most readily identifiable means of discriminating between different sorts of cycle, and are perhaps the parts which attract the most attention from the non-cycling public. The old 'roadster' used flat bars with the grips parallel with the bike, so as to provide a suitable location for the rod brake levers. This traditional kind of bar has no extra merit as far as comfort or strength are concerned. The 'all-rounder' or 'straight' bar (it is in fact bent into a very shallow W) provides a traditional sit-up-and-beg riding position, as does the mildly raised 'touring' bar. The drop handlebar, available in many patterns, looks uncomfortable, as it provides a lower, more crouched riding position, and it has a bad image amongst parents, supposedly because it is connected with the so-called 'racing' bike.

Why do nearly all cycling enthusiasts, tourists and racers alike, opt for these dropped bars? The most important fact is that they provide several different riding positions, according to whether the rider holds the tops, the brake hoods, which are usually rubber-covered for comfort, or the deepest part of the bar. In all the possible positions the angle of the rider's back is slightly different, avoiding the muscle fatigue that results from pedalling in one fixed position. Back-ache is much more likely to result during a long ride using flat bars with their immovable, unvarying riding position. Though the 'all-rounder' might look more comfortable, the more upright position is in fact less efficient, as it brings fewer muscles into play and thus places more strain on those muscles which are in use. With drop handlebars, which have a slightly longer reach, more of the body's weight is taken by the arms; consequently, as long as the forward reach is not excessive, the back is protected from jolts and bumps from the road. With the flat bars and a near vertical back, jars from bad roads are absorbed by the spine alone. Another by no means negligible asset of the dropped bar is that it allows the rider to lessen his wind resistance, and strong winds can be more exhausting for the cyclist than hill-climbs.

There are plenty of handlebar styles, but only the dropped bars provide a real and valuable choice of riding position. Bind them with two or more layers of cotton handlebar tape for extra comfort. A slight compromise for those still horrified by the idea of that apparently back-breaking bottoms-up posture is the *randonneur* style, which curls upwards before falling into a drop, thereby allowing a change of position but precluding an extra-low crouch. Special brake levers, operable from the tops of the bars as well as the bends, can be combined with this kind of bar, though the leverage from the top of the bar is not as strong as from the normal position. Whatever the style of bar, do not leave the ends uncovered. Always fit handgrips or tape, and plug the handlebar to prevent an open tube gouging into you in the event of a spill.

THE SADDLE

Last of the essential parts of the bike is that item which comes into most memorable and intimate contact with the rider, the saddle. There are two main types: sprung like a mattress or rigid; and two materials: leather or plastic. Sprung saddles are fine for short journeys, but for long runs they are liable to prove sweaty, just squashy, and no more comfortable than their spartan counterparts. They are usually fitted in conjunction with flat bars because they necessarily absorb some of the bumpiness from the road which might otherwise afflict the rider. The 'racing' or 'sports' saddle, a stiff piece of leather or plastic slung between two wire stretchers, can be extremely comfortable, in spite of its crippling appearance. Leather starts rather bony and slippy, but in use it becomes more pliable where it needs to, and it moulds itself to the individual contours of the rider after about 500 miles of riding. The plastic saddle is more instantly comfortable, theoretically more hygienic, waterproof (a soaking can ruin leather) and much lighter (by as much as a pound); but it may never achieve the ultimate, tailored comfort of the run-in leather article. Leather saddles can be bought from a few shops ready 'butchered', plied with oils and bashed around till they are pliable, but the result, though finely comfortable, is liable to stain your trousers. Perhaps the best compromise is a leather- or chamois-covered saddle on a padded plastic base, combining the instant suppleness of plastic with the 'feel' of leather.

Women cyclists, who require a shorter and wider saddle than men, have much less choice available. Apart from the unisex mattress saddle, some of which are very suitable, the only stiffer saddles seem to be plastic again. Some plastic seats can be very hard, while others have a layer of foam padding inserted between the cover and base. This looks softer but is not necessarily more comfortable, and indeed it may chafe as much as a smooth surface.

It is very difficult to generalize about saddles, but you do not have to do many miles to realize how important they are. It may take several attempts to find one that suits you. When you do find it, keep it and transfer it from bike to bike.

MUDGUARDS, LIGHTS, LUGGAGE

In most places and climates, mudguards or fenders are more or less indispensable for day-to-day riding. They pick up quite a lot of wind-resistance, however, and so it is a delight to remove them during a dry spell. It is worth checking that they can be easily removed; if the stays are attached to the frame by a bolt in a threaded brazing, rather than a nut and bolt in a smooth eye, removal is much less fiddly. Steel mudguards, unless stainless, are heavy rust-traps. In plastic, they flap around in the wind and need replacing from time to time as they split or go awry. Alloy guards are the most expensive variety and the most reliable, but they are noisy. Probably the best mudguards are plastic, as they are cheap, light and easily replaced. The short 'racing' guards serve to prevent mud and sludge from fouling the brakes, but on wet roads they will not do a thing to stop water flicking up from both wheels and soaking your legs, your backside and whoever is riding behind you.

Bicycles need to be lit at night, but the lighting systems available are not satisfactory. Battery lights fail at unexpected times, and rattle about shaking connections loose, but at least they can be instantly removed from the bike when they are not needed. They hardly provide enough light to see by, but, along with reflectors, they do serve to make the cyclist more visible to others.

French *randonneurs* strap huge multi-cell flash lights to their handlebar stems and these do throw a beam ahead. Dynamo systems cast a brighter light when the bike is moving fast and the pulley wheel is not slipping on a wet tyre, but unless they have a battery built into them they stop when you stop, which just might be unexpectedly in the middle of a poorly lit and busy road. There are also fixed hub dynamos, which avoid tyre slip, but they are heavy and mean work and revolving weight even during broad daylight, winter and summer alike. Rechargeable battery lighting systems are available, but at present these are very costly and intended only for inveterate night riders.

Luggage can be carried by bicycle in saddlebags; if these are large and heavy they should be supported by a suitable frame, or they will push the mudguard down onto the rear wheel. Panniers provide a better, lower centre of gravity. Really heavy loads should be distributed evenly between rear and front panniers, because putting all the weight on the back wheel, which takes most of the weight of the body already, is likely to result in spoke breakages or possibly the front wheel lifting at bad moments. Easily removable handlebar bags are very convenient for small odds and ends. There are bicycles intended specially for shopping which have marvellous wire baskets and detachable holdalls at the front and rear of the bike; for this purpose the small-wheeler really comes into its own. Rucksacks and shoulder bags should be avoided for anything but small journeys, primarily because they make you and the bike unstable, but also because they are almost certain to chafe. Many cyclists carry essentials in small, flat, light, envelope-like bags, called *musettes* or bonk-bags, that avoid the problems caused by heavier paraphernalia.

A bell, a pump and a good lock are all more or less essential. The simpler the bell the better. A basic one-stroke pinger will be much more reliable than an elaborate turnip-sized chime with a revolving inside. Even more basic, and perhaps more effective, you can let go a good yell. Polite cyclists can alert walkers on quiet roads by 'clacking' their brake levers.

There is a wide variety of knick-knacks available for the purpose of spoiling the essential utility and simplicity of the bike. It is pleasant to know how far you ride, but the tick-tick-tick of a mileometer can be irritating. Most kick-stands will not hold the bike up except on a perfectly flat surface when there is no wind. Speedometers put a terrific drag on the front wheel, perhaps a useful way of slowing child racers down. The need to keep the bicycle light in weight should discourage futile accessories.

So far, in this account of the modern bicycle and its component parts, we have avoided brand names and said nothing of prices. Most firms manufacturing bicycles are producing durable, serviceable and reasonably safe vehicles, with small variation in the quality of parts, but all roughly comparable in price and value according to what kind of cycle is required. You pay for extra refinement and extra quality according to your needs and the thickness of your wallet. Ultra-cheap cycles and equipment on the British market, mostly from East European firms, can be a disappointing buy; if they go wrong, which seems to be a frequent occurrence, it may be impossible to find spares, a situation exacerbated by the fact that they are often sold by mail-order firms or warehouses which have no commitment to servicing them or to importing spares. Indeed, some of the worst of them are not designed to be repaired – another reason why they are cheap.

CHILDREN'S BIKES

A word about children's bikes seems appropriate here – it is a subject that causes most parents worry and expense at some time. In this market there are bikes and 'toy bikes' and you must be sure that you can tell which is which, for price is not always a reliable guide. By the time a child is five years old you would almost certainly be better advised to avoid 'toy bikes'. They just do not last; they are difficult to repair, and sometimes difficult to adjust properly and safely. Children punish bikes in ways that adults would never dream of, and if you expect to hand the machine down to a younger brother or sister, pay a little more and get the best you can afford. Watch out for flashy finishes and inflated prices, though. Alternatively, it is a good idea to buy a quality model second hand and bring it back to running order.

You can get 'stabilizer' wheels to fit to a first two-wheeler, but if the child is big enough for the bike and you have a safe place for him to practise they are really not necessary. At first set the saddle low enough for both the child's feet to be firmly and comfortably on the ground when he is in the saddle. He should be able to 'scoot' along on it like an old hobby-horse. After a day or two at play with it you will be surprised to see him begin to master balance and then pedalling. Be on hand to give advice and assistance at this stage. If the child cannot manage like this then he may still be too young or too small for riding.

There are no hard and fast rules for distinguishing mechanical quality in children's bikes, but some of the following points and questions will help.

Be careful if you are looking at the very cheapest models, for this is obviously where the most drastic engineering economies and compromises will have been made. Ask yourself if the makers also manufacture adult bikes. The bigger names are usually reliable but not necessarily sound in every model. Look for solid construction and thorough brazing, especially at the stress points where tubes join and at the bottom bracket. Do not necessarily be put off by pressed steel front forks, they can be very strong indeed and they

will need to be to survive a kid's riding habits. Here are some further mechanical points to look for. The cranks should have cotter pins and the pedals should be replaceable. Can the saddle pin and handlebar stem be adjusted and kept firm with a stout collar clip and a big nut at each tube? These points will take a lot of wear and many changes in position. Does the bottom bracket have ball bearings in it? Do the wheel axles run on ball bearings with adjustable cones? Is the freewheel on the back axle replaceable? Ask your dealer these things. Does the cycle have inflatable tyres and at least a front brake? Are the rear seat stays fixed stoutly to the chainstays, or are they awkwardly pinched together by the bolts of the rear axle?

If the machine has some or most of these features then it is on its way to being a 'real' bike, however small it may be. Plain nylon bush bearings may be passable for play-bike wheels but they are bad news in the bottom bracket. If the cycle is one for an older child then you should look to make sure that it has an adjustable headset and bottom-bracket assembly. (This goes for adult bikes, too, especially cheap imports.) Needless to say, in all cases the brakes should be effective and adjustable, with strong mounting points and stout easily handled levers.

Bicycles in the various 'chopper' styles are still controversial and personally we favour machines with equal-sized wheels and without 'banana' seats, as they are less likely to tip over backwards. We do not like those gear levers on the cross bar either – little boys beware if you are thrown forward in a shunt. If your child is still sold on cobby wheels and motorcycle looks (despite the extra weight they entail), then there are other styles on the market which are much better balanced. Parents should not allow themselves to be talked into derailleur gears for this type of bike unless they want to become very expert in repairing and adjusting them each week. High-riser bikes are fun, of course, but bear in mind that they will be (and should be) outgrown quite quickly. The last word on the horrors of fashion is to see a lanky teenager in platform shoes trying to get a high-riser up a slight hill.

CHOOSING A BIKE

Back in the adult market and within the range of ready-made cycles, there is a certain amount of juggling around with names and colour schemes to create an apparently wide choice of bikes even when they are all produced by one firm. Raleigh, Hercules, Sun and Carlton, for example, are some of the names united in the enormous Tube Investments industrial complex. Each 'marque' produces different models, but the differences are sometimes very slight. The main distinction between some models by, say, Hercules and Raleigh seems to be that one of them has whitewall tyres and a different badge. They are all excellent machines, of course, but this kind of loyalty towards effectively extinct breeds seems peculiarly British. (In contrast, the Italian cycle-makers Bianchi produce many different bikes, but they are all called Bianchi and from the cheapest sports bike to their top-line racers they are painted the same distinctive duck-egg blue.)

The specifications in most cycle catalogues are amazingly vague anyway, and they can be changed at the drop of a hat, 'in line with our policy of continual development', as the phrase has it. Moreover, a dealer may fit different parts to standard bicycles as he thinks appropriate, or he may simply fit the wrong parts (perhaps better ones). It would thus be absurd for us to recommend the Sachs Pelican as opposed to the Sturgeon Touristique because 'we found the saddle more comfortable', as *Which?* magazine has in effect done. Your Pelican in Barnes may have a different saddle, different gear mechanism and ratio, even different wheels, if you buy it in Barnsley or if you bought it last year. Moreover, if we recommended a Crawley Sprint, routine maintenance and the occasional spare part would be a problem if your nearest stockist was sixty miles away. Do not be too encouraged, either, by a big guarantee on a bicycle's frame. It is nice to have, of course, but the frame is very unlikely to fail and lots of other things will wear out first.

By describing the relative merit and meaning of the kind of equipment readily available on ordinary bicycles, we hope to provide enough information to enable the window-shopper or bike-shop browser to understand what he sees. As for buying a bike, the same rules prevail that guide anyone trying to spend money carefully. If you are interested, send off for catalogues of bicycles and accessories. A few of the big importers (Ron Kitching or Holdsworthy, for example) produce informative booklets about their range of equipment, though much of their advice is intended for would-be racers. Poke around as many as possible of your local cycle shops – if there are any at all, that is; you may have to go quite far afield. Ask the stockist's advice. But remember that if his chief source of income is selling trikes and water pistols he may not even have heard of a cotterless chainset. Equally, if he specializes in selling very costly racing irons or hand-built frames, he cannot be expected to come alight in a debate on the relative merits of the Crawley versus the Sturgeon, with only a few pounds' difference in price and neither in his view adding up to anything more than basic transport. If he is helpful, good. But he will quite properly have a vested interest in selling you the brand of bike he stocks and is not likely to recommend the makes kept by his rivals on the other side of town.

Bicycle prices have shot up along with everything else. Most adults who last speculated about buying a bike in their early teens will be shocked when they find out the prices of even the cheaper models. It is just another sign of impinging age – they are remembering those teenage prices of one, two or three decades ago. In fact, bicycles are still good value. Compared with a toaster or a night storage heater, or even a fridge, they are complicated and finely made pieces of machinery, and, on the whole, much less shoddily finished. A good bike, well looked after, will last a lifetime. The principle of obsolescence has not yet taken over the bicycle industry, though there are signs that it is beginning to encroach on sturdily old-fashioned standards of quality and durability. The mass-produced bicycle has always cost more or less exactly the same in relation to the national average weekly wage in Britain. Since about 1920 a standard bike could be bought for one and a half times a labouring man's weekly earnings. That is to say, in 1920 about £5, in 1960 about £25, in 1970 about £45, in 1975 about £65, and in the 1980s, who knows? If you want something more than a work-horse for riding to and from your job or the shops, you pay proportionately more according to your needs – up to ten times the price of an ordinary cycle for the real thoroughbred lightweight racer.

If you do want to order a custom-built machine made to your own whims and requirements, see Appendix 2, 'Ordering a Hand-built Frame'. But you would be well advised to try a less costly bike before evacuating your bank account on a pricey racer that you might never race. To illustrate the kinds of bicycle commonly available in good cycle shops we have chosen representative models from several manufacturers' ranges, but the choice is fairly arbitrary. The descriptions may not fit exactly the specifications of the bikes pictured, but we try to indicate common variations in each 'genre' as appropriate. Once again, if you find a good dealer, stick with him; his experience and his advice can be invaluable.

FITTING

What size of cycle should you buy? Small-wheelers are usually described by the size of their wheels – 14, 16, 17, 18 or 20 inches – and they can be quickly and easily adjusted to fit almost anyone. You should be able to put a toe on the ground while in the saddle and a relaxed lean forwards should bring you to the handlebars. We would favour the large wheel diameters for comfort and ease of pedalling. The conventional diamond-frame cycle comes with wheels of 20 and 24 inches for children and 26 and, more commonly, 27 inches for adults. The size of these machines is expressed as 'frame size', which is the length of the seat tube from where the saddle pin enters it to the centre of the bottom-bracket spindle. Standard frame measurements usually run from 18 inches to 23½ inches, and a very few models go larger still. Many people ride bikes that are too big for

Folding small-wheeler. 20-inch wheels with three-speed hub gear and rear carrier. Quick-adjust levers on seat pin and handlebar stem alter the frame to fit men, women and older children. Note the special bracket to fit a dynamo on the rear seat stay. Steel wheelrims and chainset.

Gents' light roadster. An unpretentious 'commuter's' bicycle whose steel components, 'all-rounder' bars, mattress saddle and three-speed hub gears complete a typically sturdy specification. The wheels have 26-inch steel rims with plump $1\frac{3}{8}$-inch tyres.

There are many cycles of almost identical pattern on the market – reliable suburban workhorses. Despite their fondness for displaying sophisticated or sporty names like 'tourer' or 'sports lightweight', these machines are usually about the same weight as the small-wheel 'shoppers'.

Sports racer. A racing-style cycle with alloy handlebars, stem and 'GT extension' or 'safety' levers on its centre-pull brakes. The cottered chainset and wheelrims are in steel, the 'racing' saddle is in unpadded plastic. The HP tyres are 27 by $1\frac{1}{4}$ inches.

There are many models in this sector of the market and they provide an introduction to sports cycling without being in themselves true lightweights. Manufacturers ring the changes by adding or subtracting alloy equipment from the specification. You can look for a good frame with cheap equipment or vice versa, or for something in between. The better the bike the more you pay. Riders who want a true 'sports racer' look to the top end of this category or to off-the-peg machines from the specialist builders.

them, so finding the proper size is quite important. Ask your dealer if you are ever in doubt.

The easiest way to find your proper frame size is to measure your inside leg from crotch-bone to floor in your stockinged feet and subtract nine inches. If you are still in doubt over half an inch or so, then choose the smaller measurement. When adjusted, your machine should show at least two or three inches of seat pin – any less and it is probably too big for you. The measurement is not absolutely critical for everyday riding, of course, but if you are buying a bike you might as well get it right. Advice on how to find a comfortable

Ladies' 'mixte' light roadster. The stem, 'all-rounder' handlebars and centre-pull brakes are all alloy; the HP wheels and cottered chainset are steel. The five-speed derailleur is operated by a lever on the stem. Chain-guard, prop-stand, padded saddle and rear carrier are fitted as standard.

Quality ten-speed tourer. Reynolds 531 plain-gauge frame. Alloy HP rims and large-flange quick-release hubs. Cotterless alloy chainset with 36 and 48 chainwheels and rear sprockets up to 32 teeth give a very wide range of gears. Note the brazed-on fittings for lights, the cable stops on the top tubes and the rear bridge for centre-pull brakes. Duo brake levers, leather saddle, custom rear carrier, tough plastic mudguards, toe-clips, alloy pedals and pump complete a very good touring specification.

rofessional road-racing bicycle. The frame is Reynolds 531 double-butted throughout. The wheels are all-alloy with large-flange hubs, sprint rims and 8½-ounce tubular tyres. The free wheel has six cogs to give a possible twelve speeds. The bottle-cage is bolted to brazed-on eyes on the down tube. The plastic saddle is lightly padded and covered with suede. The frame angles may be as steep as 75 degrees and the wheelbase is 39¼ inches.

Most of the alloy equipment on this particular example comes from the Campagnolo Super Record range which uses titanium for extra lightness, and this alone makes the bike very costly indeed. An excellent racing machine can still be assembled by choosing much less expensive components. The most important items to spend your money on to begin with are a good frame and wheels.

and efficient riding position can be found in Appendix 1, 'Maintenance, Comfort, and Safety on the Road'.

After so many details and the naming of so many parts, let Samuel Beckett's Molloy have the last word; he is, after all, an experienced rider, although he has one or two unique fitting problems himself:

> . . . I was no mean cyclist at that period. This is how I went about it. I fastened my crutches to the cross-bar, one on either side, I propped the foot of my stiff leg (I forget which) on the projecting front axle, and I pedalled with the other. It was a chainless bicycle, with a free-wheel, if such a bicycle exists. Dear bicycle, I shall not call you bike, you were green, like so many of

your generation. I don't know why. It is a pleasure to meet it again. To describe it at length would be a pleasure. It had a little red horn instead of the bell fashionable in your days. To blow this horn was for me a real pleasure, almost a vice. I will go further and declare that if I were obliged to record, in a roll of honour, those activities which in the course of my interminable existence have given me only a mild pain in the balls, the blowing of a rubber horn – toot! – would figure among the first . . . What a rest to speak of bicycles and horns.

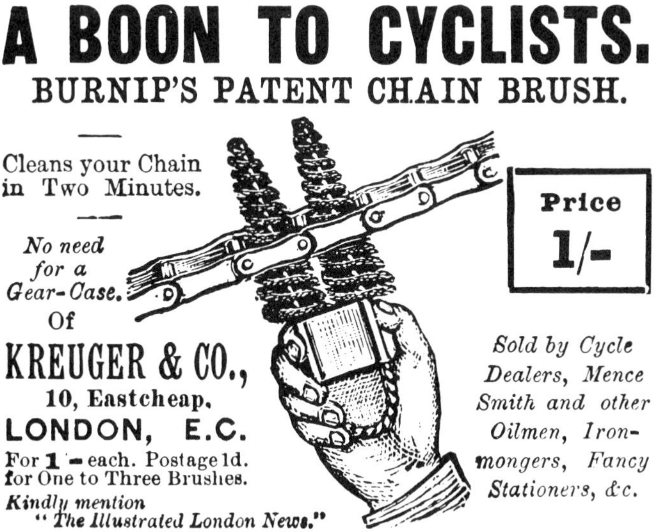

THE LINE OF BEAUTY.

Athletic. "DON'T *YOU* BICYCLE?"

Æsthetic. "ER—NO. IT DEVELOPES THE CALVES OF THE LEGS SO! MAKES 'EM STICK OUT, YOU KNOW! SO COARSE! POSITIVE DEFORMITY!!"

Punch, 1879

3 The Aesthetics of it All

How puzzled some of those with whom we have had pleasant journeys on the velocipede looked when they first mounted the two-wheel, 'dangerous looking', hoop-like machine. 'Oh! Hold me up; I shall never master it, etc., etc.,' were the exclamations; but a day or two afterwards these very men rushed in on us with some such phrase, 'Oh! what jolly fun; I have come from . . ., and never tumbled once!' These very men have often afterwards described the easy and rapid locomotion as an enjoyment which was *positively intense*.

The Velocipede: Its History and Practical Hints How to Use It,
by an experienced velocipedist, London, 1869

At first sight the bicycle and aesthetics may hardly seem to go together. To most people a bike is about as art-full as a boot, along with the bedsteads and rusty kettles that comic fishermen find in canals. The awkward ghost of that hatless post-war Englishman is still riding. Let us exorcise him once and for all by considering what might possibly be called beautiful about the bicycle in the eyes of the engineer, the philosopher and the athlete. A good common base for all these points of view would be what Lewis Mumford in *Art and Technics* calls 'significant form'. So let us begin by considering the bicycle as what he calls a 'type object':

. . . once established and perfected, type objects should have a long period of use. No essential improvement in the safety pin has been made since the bronze age. In weaving there has been no essential modification of the loom for over a century. And what is true for machines holds good in no small degree for their pro-

ducts. When the typical form has been achieved, the sooner the machine retreats into the background and becomes a discreetly silent fixture the better. This again flies in the face of most contemporary beliefs. At present, half our gains in technical efficiency are nullified by the annual custom of restyling. Extraordinary ingenuity is exercised by publicity directors and industrial designers in making models that have undergone no essential change look as if they had. In order to hasten style obsolescence, they introduce fake variety in departments where it is irrelevant – not in the interest of order, efficiency, technical perfection, but in the interest of profit and prestige, two very secondary and usually sordid human motives. Instead of lengthening the life of the product and lowering the cost to the user, they raise the cost to the user by shortening the life of the product and causing him to be conscious of mere stylistic tricks that are without any kind of human significance or value. This perversion of technics in our time naturally saps the vitality of real art; first by destroying any sound basis for discrimination and then by taking energy and attention away from those aspects of human experience in which the unique and the personal are supremely important.

The concept of the Safety bicycle dates from 1885, and in material and design most modern roadsters go back to 1900. But the bicycle *has* changed, if very subtly, and these changes continue to take place from year to year, from racing season to racing season. How close the bicycle has come to technical perfection and whether it has an aesthetic value or not will become clearer as we go. Indeed, the question may well be whether 'technical perfection' can ever be achieved at all. But let us start by looking at the cycle's status as a social machine. If there is an aesthetic here it is surely an important one, concerning, as it does, the relationship between the environment and our bodies.

Unlike the car, so often used only as a motorized shopping basket, the cycle maintains an appropriate balance between the individual, his freedom in relation to that of others and his power of movement. At the very most, four cyclists need half the amount of space required by a car moving through the High Street, and that car probably carries only one person. Those four riders are silent and non-polluting, and the manufacture of all their machines has used only $\frac{1}{250}$ of the energy needed to make the car in front of them. It is this human and responsible dimension that makes the cycle a potentially vital and liberating factor in our understanding of urban life and planning. This is part of what Ivan Illich means when he calls it a 'convivial tool' and this 'conviviality' is shared by all cycles from the small-wheel 'shopper' to a racing machine to the grocer's humble work-horse. This chapter, however, will concern itself mainly with the pleasures of the lightweight machine, since it is here that the finest state of the art may be found.

If we are talking about tools, then the bicycle is an efficient tool, as S. S. Wilson explained in an article for the *Scientific American*. Wilson compared the mechanical efficiency of the bicycle with that of a man walking and found that the cyclist needs only one fifth of the walker's expenditure of energy to travel three or four times faster. In fact this improves on nature to such an extent that it makes the cyclist the most efficient of all moving animals and machines. Wilson showed this on a graph. As a matter of interest, do notice the huge energy consumption of such exotic flyers as the hummingbird and the helicopter compared to that of the humble pigeon, an economical and versatile bird that can be bred for racing or eaten in pies.

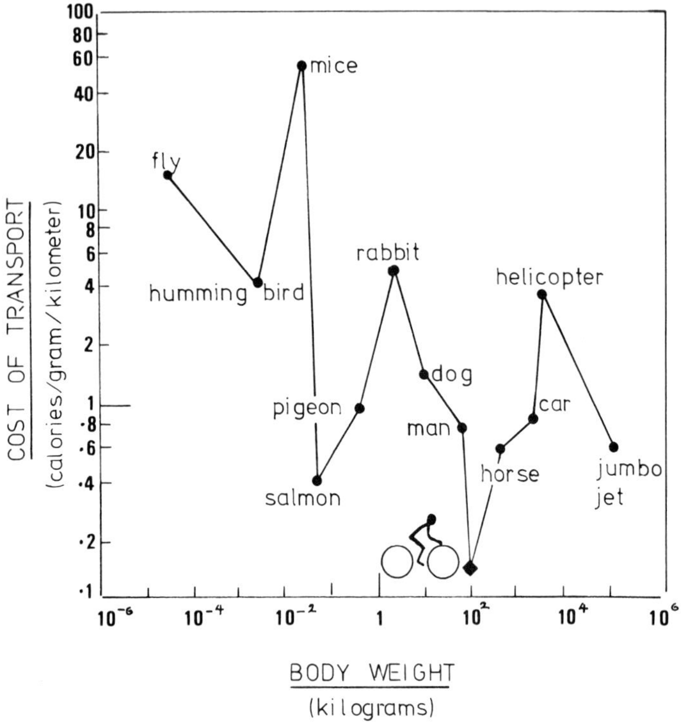

Fig. 9. Based on a graph by V. A. Tucker of Duke University, North Carolina.

An enthusiastic and entertaining article by Dr R. E. Williams in the *British Medical Journal* expressed the same information in physical and statistical terms:

ENERGY CONSUMPTION

Our patients, the long-suffering and generally under-exercised British public, continue to show such massive unawareness of the many advantages of the bicycle that it becomes a therapeutic duty to contrast for them this clean, quick, quiet, and civilized machine with that oxygen-eating, air-defiling dissipator of energy – my faithful motor car. I say faithful because anything less would be churlish, and it is with real reluctance that I expose to public gaze the imperfections of a family friend. But already the clouds are gathering round the internal combustion engine as we have used it, and we are being forced to consider less wasteful ways of getting ourselves about. For a car uses only 20% of the combustible energy of its fuel in moving forward: 4% goes to essentials such as transmission, dynamo, fan, and water pump, but over 75% is lost as heat – 40% through the exhaust and most of the remaining 35% by conduction and convection through something misleadingly called a radiator (the only thing it does hardly at all). The cyclist is altogether more temperate. Indeed from the standpoint of physics he is not a heat engine at all, but a constant-temperature energy converter more analogous to a fuel cell; and his contribution to the increasing entropy of our solar system is commendably small. His expenditure of energy in terms of his car's use of petrol is equivalent of 1,500 miles to the gallon (536 km/l), and, which puts him even more on the side of the angels, he leaves the air cleaner than he finds it, thanks to the trapping of inspired particulate debris by the mucous sheet covering his bronchial and branchiolar epithelium.

ECONOMY

In the past three years I have travelled 6,000 miles (9,660 km) on my bicycle which otherwise I should have done in my car – an average of eight miles (12.9 km) per working day. The saving in time has been more hours than I can count, in petrol more than 240 gallons (1,090 litres), and to the atmosphere about 2,000,000 litres of oxygen, 1,300,000 litres of carbon dioxide, and 300,000 litres of carbon monoxide, at normal temperature and pressure. Now 2,000,000 litres of oxygen sounds a lot but you may well say, 'Why the fuss? The earth is large and its atmosphere miles deep; 2,000,000 litres is only 70,600 cubic feet (1,977 m^3) of pure oxygen which is not much more than twice the capacity of the Long Room at Lord's (32,400 cubic feet; 908 m^3). All you have saved in three years' cycling is the oxygen content of the air of 11 Long Rooms.' And you would be right. But these gas volumes derive from the very modest mileage of a single motor car. Compounded for a city the amounts are more disturbing. At continental levels they require at least some thought.

To look at this human scale another way, almost every dimension of the modern lightweight cycle is proportioned to the body of the rider. The size of the frame is a function of his height (nearly one third of it) and the width of the handlebars should be determined by the width of his shoulders. Chapter 2 and Appendix 1 outline the various ways in which a bike can be adjusted to its rider by using his own body as the scale, and a well-set-up machine quite quickly becomes an extension of his limbs and his physical potential. The cyclist's weight is distributed between the saddle and the bars, with his back at the correct angle for delivering power through the whole length of the thigh, using only the strongest muscles. This position also helps to spread the shocks from a bumpy road, distributes the weight more evenly over the front and rear wheels and holds the rib cage in a relaxed position for full and deep breathing. The toe-clips on the pedals bind the rider's feet to the cranks so that there is not an ounce of lost force as his legs pull upwards and thrust downwards in a smooth rotary motion in which each leg is aided by the thrust of the other on the opposite crank. Even an average machine will add up to only one sixth of the rider's weight, and this lightness, the low rolling resistance of high-pressure tyres and the use of gears all make for rapid and easy movement.

In fact the lightweight two-wheeler is so efficient as a means of transmitting muscle energy that it is difficult for a fit rider on level ground to raise his pulse rate much above 120 a minute. At every turn the bicycle is characterized by this balance between mechanical efficiency and human potential. Air resistance, for example, accounts for most of the work required to pedal at speeds over twenty miles per hour. But the very same air flow carries away the heat generated by the rider's efforts and this natural cooling effect is vitally important. Indeed, as Whitt and Wilson describe in *Bicycling Science*, experiments carried out on a static ergometer in the laboratory suggest that riders can sustain 'speeds' of twenty-five miles per hour for only ten minutes or so before they overheat and 'blow up'. Yet on the open road, fast time triallists can keep this same pace even for as long as four hours.

On a good day, when the road is right and you are feeling strong, the stiff and yet responsive frame feels like an inseparable part of your legs and shoulders as you power through a corner. Your lungs breathe in whole mountains, streams, trees, sheep, as you fly across the countryside, drunk on air and motion. You have become the 'Cartesian centaur', as envisioned with cerebral passion by the literary critic Hugh Kenner:

Consider the cyclist as he passes, the supreme specialist, transfiguring that act of moving from place to place which is itself the sentient body's supreme specialty. He is the term of locomotive evolution from slugs and creeping things. Could Gulliver have seen this phenomenon he would have turned aside from the Houyhnhnms, and Plato have reconsidered the possibility of incarnating an idea. Here all rationalist metaphysics terminates (as he pedals by, reciprocating motion steadily converted into rotary). The combination is impervious to Freud, and would have been of no evident use to Shakespeare. This glorified body is the supreme Cartesian achievement, a product of the pure intelligence, which had preceded it in time and now dominates it in function. It is neither generated nor (with reasonable care) corrupted. Here Euclid achieves mobility: circle, triangle, rhombus, the clear and distinct patterns of Cartesian knowledge. Here gyroscopic stability vies for attention with the ancient paradox of the still point and the rim. (He pedals with impenetrable dignity, the sitting posture combined with the walking, *sedendo et ambulando*, philosopher-king.) To consider the endless perfection of the chain, the links forever settling about the cogs, is a perpetual pleasure; to reflect that a specified link is alternately stationary with respect to the sprocket, then in motion with respect to the same sprocket, without hiatus between these conditions, is to entertain the sort of soothing mystery which . . . you can study all your life and never understand. The wheels are a miracle; the contraption moves on air, sustained by a network of wires in tension not against gravity but against one another . . . Here is the fixation of childhood dream, here is the fulfilment of young manhood. All human faculties are called into play, and all human muscles except perhaps the auricular. Thus is fulfilled the serpent's promise to Eve, *et eritis sicut dii* [and you shall be like gods]; and it is right that there should ride about France as these words are written, subject to Mr Beckett's intermittent attention, a veteran racing cyclist, bald, a 'stayer', recurrent placeman in town-to-town and national championships, Christian name elusive, surname Godeau, pronounced, of course, no differently from Godot.

Samuel Beckett: A Critical Study

Yet, *pace* Hugh Kenner, the serpent and Descartes, the bicycle is not totally rational. It would be less interesting, less engaging, less 'human', if it were. Several elements in its design are the way they are simply because that is how they developed from earlier versions. Like our own bodies, the cycle carries the evidence of its evolution within it.

'. . . *sedendo et ambulando*' in the Indian monsoon.

For example, bikes do not *have* to be guided by way of the steering column running from the forks to the handlebars through a head tube in the frame. Other methods have been tried and some of them are mechanically more efficient. With the present arrangement road shocks tend to be translated into forces which can very slightly distort the steering geometry (see Figure 10). This is why good-quality, well-adjusted bearings are important at this point (the 'headset'). Of course, the forces involved are well within the safety limits of the materials, so this design has been the most common one in the past and will probably continue to be so for as long as it is cheaper and lighter to produce than a more sophisticated geometry. Notice, too, that the chainwheel of nearly all bicycles has been on the right side of the frame. (This was standardized in 1898.) This can be explained by the fact that most of us are right-handed, and that the rear sprocket or freewheel is fitted with the usual right-hand thread. If further explanation is required, it seems likely that our maximum muscular effort can be brought to bear on this side and so this is the best place to position what is, in effect, a lever.

Thus in these and other ways the cycle represents a subtle compromise between the demands of engineering and those of our

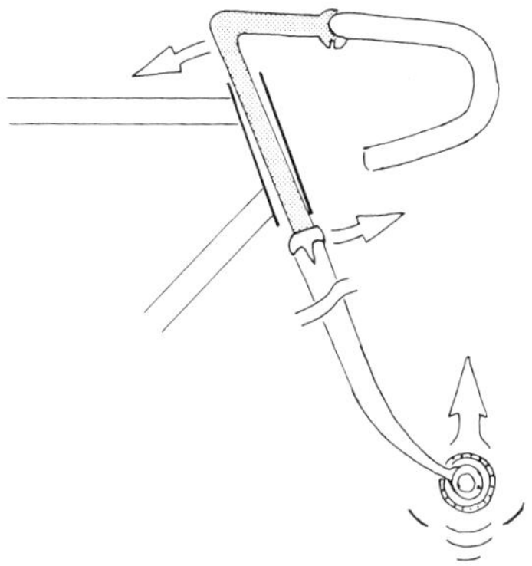

Fig. 10

bodies. Indeed, its evolution has been more intimately bound up with our own physical capacities and dimensions than that of any other machine one can think of. The best example of this inter-relationship can be found in the frame itself. The skeleton of the bike is composed of a set of tubes built into triangles, and this pattern is especially suitable for light structures which have to resist strong loads. Even a thin tube of circular cross-section will resist powerful forces trying to bend, twist, stretch or compress it. The triangulation of the frame is a way of reinforcing these qualities. Thus, in theory at least, the rotary motion of the crank is supported at the bottom of the main frame tubes somewhat as in Figure 11. On a racing cycle during a hard sprint the force acting on each pedal in turn, driven through the leverage of the cranks and thus across the bottom bracket, can reach some 350 pounds, twice the rider's weight. It is very important that the frame be stiff enough

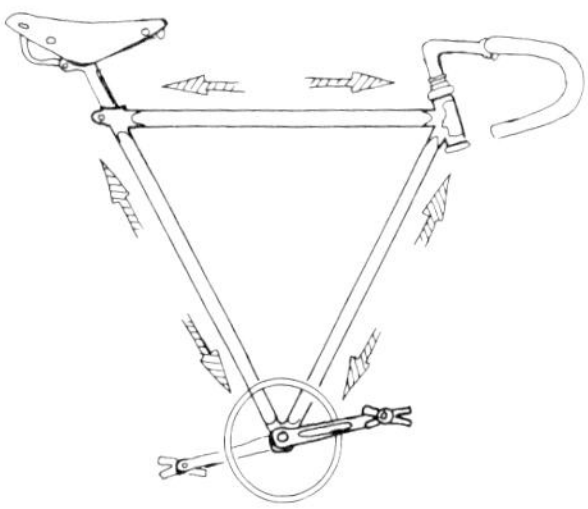

Fig. 11

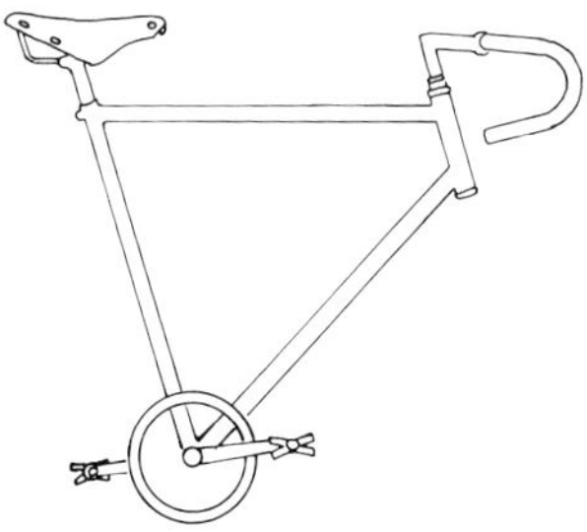

Fig. 12

to resist these huge bending forces and to transmit them instead to the rear wheel. At the same time, however, a frame which was a hundred per cent rigid (if it could be built and linked to equally rigid wheels) would lack an important element of responsiveness and probably provide a very uncomfortable ride.

In practice, of course, the cycle frame is not absolutely rigid, nor is it a perfectly triangulated construction. This is mainly because of the need for a head tube for the steering column. Thus the frame becomes a quadrilateral figure which usually has two sides nearly equal in length and two parallel, as in Figure 12. When we add the stays to hold the rear wheel (these do not have to be quite so strong) we have the conventional diamond frame as in Figure 13.

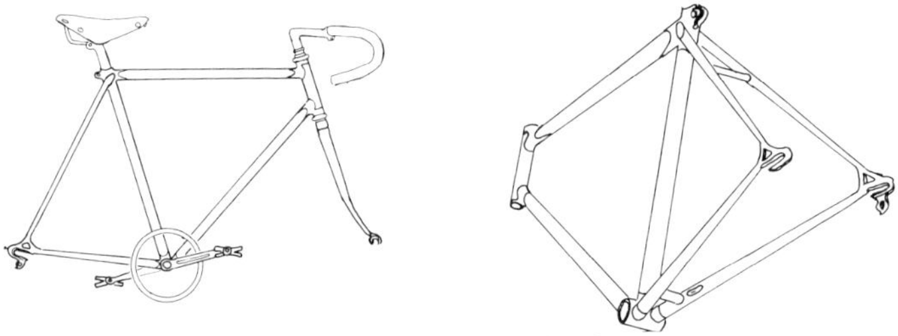

Fig. 13. The diamond frame and the rear triangles.

Even though it is an engineering compromise, this pattern is stronger than any bridge, aeroplane or motorcar because it is capable of carrying about fifty times its own weight, and it will resist a load of two tons applied vertically at the top of the seat tube.

As a matter of fact, a more properly geometrical frame was produced in the 1890s by using small tubes triangulated in pairs on the same principle as modern racing-car 'space-frames'. This machine was called the Dursley–Pedersen, weighted only 23 pounds and still has enthusiastic supporters today. It had many original features, including better steering geometry and a most unusual saddle (see Figure 14). Each frame member consists of two tubes in a triangle. The cantilever construction means that these light tubes take compression stresses only and the triangulation across the bottom bracket controls the lateral distortions produced by pedalling. Notice the wire rigging to the rear hub and the unique 'hammock' saddle. The steering head and forks show how different the design is from the modern head-tube compromise.

Fig. 14

Pedersen was a Danish engineer living in England; his machine was patented in 1893 and manufactured from 1897 to 1902 in the Humber Works at Beeston and then at the Lister works in Dursley until 1914: twenty-one years of quality. In 1897 the Pedersen cost £25, although later models were marketed more cheaply. Enamel, nickel-plated or copper-burnished finishes could be specified and a ladies' model was also made. No other bicycle has been so scientifically designed from first principles. A modern replica using the latest materials (but probably without that odd saddle) would be a most interesting project.

The exact proportions of the diamond frame are important because they govern the kind of ride the cyclist will experience. Thus for touring the rider will want a longer wheelbase, shallower angles at the head and seat tubes and a larger off-set on the front fork. These qualities (greatly exaggerated in Figure 15) will produce a very stable ride, a machine that will soak up rough surfaces and practically steer itself, a bike to sit on all day without discomfort. The saddle would be set lower than a full racing position and the raised 'Randonneur' handlebars make for a leisurely semi-upright position. The low bottom bracket ensures stability, especially when carrying luggage, and the small chainwheel and the wide range of large rear sprockets will cope comfortably with even the steepest hills.

Fig. 15

If, on the other hand, only the smallest changes are made to those triangles – a matter, say, of three inches and three degrees – then the entire character of the machine is transformed. Steep angles at seat tube and head tube, a short wheelbase and close-coupled wheels will produce a cycle that is highly responsive in sprints and hill-climbing, but one which needs skilful riding and concentration. It will be a 'nervous' mount. The handlebars are set low and the saddle high to produce maximum power with a minimum of wind resistance. The high bottom bracket allows the cycle to be cornered at steep angles without the pedals grounding, and the large chainwheel and the close-ratio sprockets at the rear are designed for riding with a sustained and fast action, producing maximum power at all times (see Figure 16).

In fact, a difference of only a few inches and a few degrees gives us a fully fledged racing machine designed for an athlete. Even a complete cycling novice would notice an immediate and obvious difference in 'feel' if he were to ride these two machines.

Most cycles, of course, compromise between these two extremes but whatever the purpose for which the bike is designed there is an attractive and, as we have seen, a fundamental reason for its proportions. The lines of a bicycle dictate its nature absolutely; its looks and its function are one. Thus 'good taste' in the finish and accessories of a bicycle consists of letting this unity speak for itself as clearly as possible.

Fig. 16

It is largely because of this integrity that the bicycle has changed so little in principle since John Kemp Starley's Rover Safety of 1885. Hundreds of different accessories and inventions have been applied to it, but in the end they have fallen by the wayside, sometimes all too literally. The really important developments were all made early: the tangent spoked wheel in 1874, the perfection of the ball-bearing race in 1877, the bush roller chain in 1880 and the derailleur gear in 1899. Indeed, it was the factory production of the bicycle which stimulated the development and wider application of these and many other important inventions which were to be fundamental to the rise of the motorcar.

If the bicycle has changed little since Starley's day, it might equally well be said that its development has stagnated, and critics would point to the cycle brake, the derailleur gear and the problem of punctures to support their case. The rim brake operates at exactly the point where it is most likely to be affected by water, and although alloy rims and studded brake blocks go some way towards solving the problem they are still far from perfect. Grown engineers have been known to blanch when they are shown how the derailleur literally derails the chain from one whirling sprocket to another one even further offset from the driving chainwheel. Nevertheless the brake and the derailleur gear, like the diamond frame, do represent a fairly highly developed compromise. On the other hand the puncture, like bad weather, is always with us. (They even seem to happen together.)

Hub brakes and even disc brakes are available for the cycle, but their increased mechanical efficiency carries with it a considerable penalty in weight. And weight is the cyclist's enemy. Further, the more efficient the brake, the greater the stress undergone by the wheel and tyre on stopping, and, since so little of a lightweight tyre is in contact with the ground anyway, the whole system would have to be made stronger and heavier. This would in turn increase the mass and the inertia to be overcome by the brakes in the first place. And so it goes on.

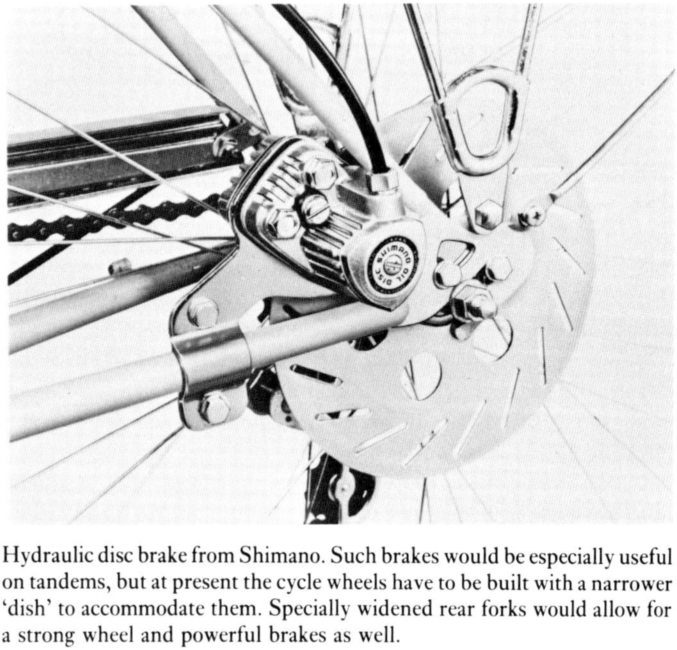

Hydraulic disc brake from Shimano. Such brakes would be especially useful on tandems, but at present the cycle wheels have to be built with a narrower 'dish' to accommodate them. Specially widened rear forks would allow for a strong wheel and powerful brakes as well.

The same principle and penalties apply to the drive train. A well-maintained chain loses only 1.5 per cent to friction, compared with power losses of up to 15 per cent for a car gear-box, which has to turn lubricating oil, idler shafts and all those cogs. Pioneer lady cyclists who bought shaft-drive machines were familiar with this effect. They found that they had to work harder at the pedals, though they were freed from the worry that their skirts would be caught in a messy chain. Being a lady on a bicycle was not easy in those days; even so, the shaft-drive machines did not prosper.

The cycle hub gear has less machinery than a car gear-box and under power its friction losses are only about 4 to 6 per cent, except in direct drive, where they are minimal. This compares surprisingly well with the derailleur gear, which loses as much as 5 per cent to friction. Derailleurs also require fine adjustment and regular maintenance, and they are positively delicate compared to the grit-proof sturdiness of the hub gear. As late as the 1930s Tour de France riders scorned the derailleur as a tourists' device too frail for their hard conditions. They had freewheel sprockets at each side of their back hubs and they changed gear in preparation for the climbs by stopping and turning the wheel around. Nevertheless, despite all its drawbacks the derailleur has greatly improved in recent years and it seems unlikely to be bettered as a racing gear. It is lighter than a hub system, it can give five to twelve speeds – or even fifteen for the keen tourist – and the rider can alter his gear ratios fairly easily by replacing sprockets on the freewheel. These features give it the edge for most serious cyclists. As a Frenchman is reputed to have said with admirable pragmatism: '*C'est brutal, mais ça marche!*'

Perhaps the most welcome advance of all would be a puncture-proof tyre. The racing man would be even more enthusiastic about it than the tourist, since many competitions are won or lost by the time it takes to change a wheel. Modern rubber is very hard-wearing, of course, and puncture-resistant latex tubes are available, as well as a liquid sealer to be injected through the valve. But it is still possible with careless or emergency braking to rip a hole clean through a light cover (a stray sheep proved this to us once). A solid tyre may yet be invented which is as light, resilient and cool-running as the pneumatic type, and it is surprising that some sort of air-expanded rubber or plastic core has not been marketed. One of the drawbacks is that a 'solid' or foam-filled tyre will not absorb road shocks as well as a pneumatic tube does, and thus it is more likely to dent the wheel rim when it meets an unexpected pot-hole. Nevertheless, a substance called 'monofil' which promises well in this direction has been used in America and it may not be too long before the tyre problem is solved. But of course the next question is: how much will it cost?

These are some of the factors that suggest that the bicycle has not stagnated but has reached a steady state between conservatism and innovation in which almost all its parts have struck a subtle compromise between strength, efficiency, economy and lightness. Yet this is a very delicate state. Attach a motor to the cycle frame, for example, and its balanced design is instantly upset. The chain becomes comparatively weak and inefficient where before it was more than adequate for human muscles (since cycle chains are made to aircraft control specifications). The gears, brakes and wheels simply collapse. The inventor will have to look for different solutions or stronger materials. He will, in fact, go on to produce the motorbike, a machine which is itself a nicely balanced (but less subtle) collection of compromises.

The history of transport shows that every increase in motive power has demanded corresponding developments in all other aspects of the vehicle. But the cycle alone is powered by human energy and this, for all practical purposes, is a constant. Therefore, given a trained rider, increases in cycling performance can only be gained by the refinement of designs and by the lightening of components. It is not just a question of drilling holes in everything, however, for this refinement takes in some very sophisticated technology indeed.

Thus we have the fascinating case of a machine whose basic engineering goes back to 1885 but some of whose parts may be made in aerospace plastics and alloys originally developed to put men on the moon. Cycle frames are currently being made in aluminium, titanium and carbon fibre, but the exotic materials involved and the considerable problems of building them with brazed, welded or screwed and glued joints mean that they are very expensive indeed.

Most cycle parts are, however, mass-produced, and this makes excellent sense because it allows high-quality steel tube or alloy castings to be produced very economically by large foundries which may make many other products as well. Thus the individual builder can use material of uniform quality made to specifications far higher than any he could himself attain. The assembly of a bicycle is not in itself a difficult process and thousands of machines are mass-produced on production lines and give excellent service. Many hand builders have their frames made by several men, each of whom works on one special part of the job. This speeds up production, maintains a high level of finish and allows for more individual features to be accommodated on each bike. But some of the best machines are still made by one man whose skill and judgement create an individually balanced and very highly finished piece of work.

Surely, at some mid-point between that skilled individual and the mass impersonality of the assembly line, there exists a model of how we should plan our relationship with technology? The whole of culture and society stands to benefit from a more creative and responsible relationship between who we are and what we do. SAAB of Sweden have made excellent advances in this field at their factory in Södertälje, where engines are finally assembled by groups of only three workers who handle 'their' engines from first to last. This approach greatly improves work conditions as well as the quality of the product, and in 1973 a work force of only three hundred was making 110,000 engines a year. The employees of the small cycle-builder were enjoying this kind of relationship with their work long before industrial theorists began to worry about 'quality of life' on the shop floor.

The quality of a hand-built bike can often be judged by the finish on the lugs of the frame, and it is here, too, that the workman can introduce an element of decoration or self-expression in his metal-work. Lugs are necessary because frame tubes cannot be welded at high temperatures without causing crystalline changes in the steel and subsequent weakening of the joint. Instead the tubes are heated relatively gently and brazed together with a molten charge of copper–zinc (brass) alloys between the tubes and the lugs, which work like sockets joining them together. These lugs are cast or pressed in steel to whatever angle the frame requires. They serve to strengthen the joint between the tubes (which are mitred together inside each lug) and they also help to spread the loads which accumulate at the apex of the frame triangles. The edges of each lug are filed down to make an easier gradation between the reinforced part and the rest of the tube and also to help spread the heat from the brazing process more evenly.

This work requires a lot of preparation and hand finishing and so it is here that some custom builders choose to show their skills by producing cut-out patterns and filigree edges. This decoration is not essential, but it is not entirely gratuitous either, and builders explain that the cut-outs help the workmen see whether the brass has penetrated the whole joint and that the lengthy curled edge also helps spread the changes in stress from reinforced to non-reinforced parts. This is true, but the same engineering requirements are equally well met by the plain long-point lug, even if it does not look quite so flamboyant.

By way of comparison, consider the fine engraving and inlay traditionally applied to sporting guns. Here, too, function, engineering, craftsmanship and decoration come together happily – if for a more macabre purpose. On the other hand, it can be claimed that the structural function of the lug is best expressed in its plainest form, without decoration applied extraneously like the wilder ornamentations on Victorian sewing machines. Lewis Mumford maintains that 'the canons of machine art are precision, economy, slickness, restriction to the essential.' Personally we favour this point of view, and some cyclists can scarcely abide even a transfer on their mounts. But fancy

Fancy lugs from Condor and the cut-out bottom bracket on a Dawes frame.

lugs are usually confined to the head tube only and, given the bicycle's still basic austerity, who could condemn this one small flight into rococo?

The quest for lightness and the skill of the builder reach their highest point in the construction of specialist frames for record attempts and time trialling. In these events the bike will not be subjected to the punishment that a racing machine has to take in the hurly burly of a mass of riders on the road. A super lightweight can run on fewer spokes (say, 24 instead of the more usual 36 or 32 to each wheel), the bottom bracket may be cut out (the bearings protected by a sleeve of plastic), the chain stays may be slotted, seat pin and handlebar stem may be fluted and allen bolts in Ergal may be used, all to save a few precious ounces. This is a specialist machine, however, built to be handled carefully and smoothly by an

Lightness and lessness: a Zeus chainwheel, a slotted chain and the Shimano 10-mm pitch power train.

experienced rider. Nothing could be worse or more dangerous than an inexperienced young rider with a power drill trying to 'save weight' with a hole here and there.

For all the maker's skill with fancy lugwork, the cycle cannot rightly be called 'a work of art' in the prestigious context of modern art galleries. Nevertheless, its aesthetic involves values and understandings which make some exhibits in these galleries look pretentious and ill-made. (Isn't there a South Sea Island language which has no word for art 'because we make all our things as well as we can'?) Furthermore, the cycle is an entertaining machine because, unlike almost all our other industrial products, the more you pay the less you seem to get, and this is revolutionary in its implications. Another intriguing thing about it is that its virtues are so abstruse as to be almost invisible to the novice and yet they are there in front of him all the time if he knows what to look for. For example, Italian Campagnolo hubs are jewels of light alloy with their barrels and flanges machined from solid and their (invisible) internal bearing surfaces of highly polished steel. A well-trued cycle wheel will revolve so smoothly when it is suspended that it should always come to a stop with the tyre valve at the bottom. Even its spokes are thinner in the middle and thicker at each end (double-butted) to give the best in both lightness and strength. Under tension these spokes are a shimmer of extraordinarily complex forces and to compare a cycle wheel to a car wheel is rather like comparing the airy grace of a suspension bridge to a plank across a ditch. Not that the same forces are absent in the car wheel, it is just that they are supremely seen to be there in the cycle wheel, demonstrated in every line.

'Suspension bridge' is right, in fact, for the machine and the rider himself are indeed truly suspended in air by the spokes from the top of the rim. If these spokes are built radially – taking the shortest distance from the centre to the rim – then the wheel will be very stiff and give a rigid ride. If, however, tangent lacing is used, in which each spoke crosses over one, two or the more usual three or four of its neighbours before it reaches the rim, then longer spokes are needed and the wheel will give a more efficient 'pull' and also a more comfortable ride. Radial or one-cross spoking is sometimes used by time triallists, who may have only twenty-four spokes to each wheel. Consider finally the subtleties of the rear wheel, which of course has to be 'dished', to keep the rim on the centre line of the axle and to allow for the freewheel cluster. To help offset the differences in tension created by different-length spokes, a rear wheel can be laced four-cross on the sprocket side and three-cross on the other.

In every other mechanical aspect the bicycle comes to us equally naked – wheels, pedals, chain, crank and forks demonstrate their purpose and only their purpose, with scarcely an ounce of surplus matter. The priorities of lightness and strength do not leave room for 'optional extras', and the philosophy of 'lessness' creates a purity and an austerity that has almost disappeared from other manufactured goods. The cycle has the beauty of gliders, skis or racing sailing boats because it is so supremely and visibly at one with the natural forces which define it and which are, in their turn, overcome by it.

Lessness on a time-trial machine weighing only 19 pounds and fitted with very costly Campagnolo titanium equipment.

This integrity is the essence of what Robert Pirsig calls 'Quality' in his book *Zen and the Art of Motorcycle Maintenance*. Here is his version of the state achieved by our Cartesian centaur:

At the moment of pure quality, subject and object are identical. This is the *tat tvam asi* truth of the Upanishads but it's also reflected in modern street argot. 'Getting with it', 'digging it', 'grooving on it' are all slang reflections of this identity. It is this identity that is the basis of craftsmanship in all the technical arts. And it is this identity that modern, dualistically conceived technology lacks. The creator of it feels no particular sense of identity with it. The owner of it feels no particular sense of identity with it. Hence, by Phaedrus' definition, it has no Quality.

That wall in Korea that Phaedrus saw was an act of technology. It was beautiful, but not because of any masterful intellectual planning or any scientific supervision of the job, or any added expenditures to 'stylize' it. It was beautiful because the people who worked on it had a way of looking at things that made them do it right unselfconsciously. They didn't separate themselves from the work in such a way as to do it wrong. There is the center of the whole solution.

The way to solve the conflict between human values and technological needs is not to run away from technology. That's impossible. The way to resolve the conflict is to break down the barriers of dualistic thought that prevent a real understanding of what technology is – not an exploitation of nature, but a fusion of nature and the human spirit into a new kind of creation that transcends both. When this transcendence occurs in such events as the first airplane across the ocean or the first footstep on the moon, a kind of public recognition of the transcendence should also occur at the individual level, on a personal level, on a personal basis, in one's own life, in a less dramatic way.

It does not have to be a journey to the moon, for every cyclist has at some time found something of Pirsig's state of identity between the world and his machine.

Such moments of transcendence where subject and object become one are common in various kinds of Yoga and meditation practice, but recent writers have become increasingly interested in similar states occurring in sports. The balance of the Martial Arts, the perfect swing of the golfer or the 'unwilled' shot from the archer all lend themselves to Zen-like techniques and exposition. Cycling also shares this potential and, indeed, many racing cyclists are familiar with Yoga exercises. The essence of these transcendental states is deep and full breathing combined with an inner calmness and detachment. What could be more appropriate for the cyclist – especially for the time triallist who rides alone against the clock in what the British like to call 'the race of truth'? Certainly it is exciting and greatly refreshing to open the mind only to 'stillness' on the

HUMBER & CO., LIMITED.

HOLDS THE WORLD'S RECORD.

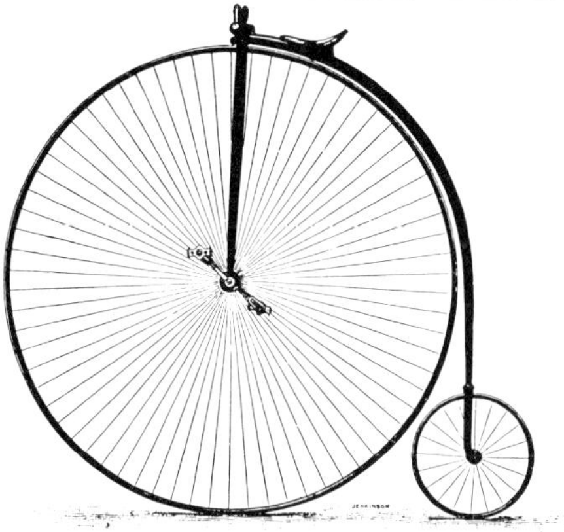

ONE MILE 2 min. 31 4-5ths Sec.

HUMBER No. 3 RACER BICYCLE

(Beeston Make).

Is the FASTEST BICYCLE in the world. Weight is saved in every possible way without detracting from strength and rigidity, and the wonderful successes achieved by racing men on this Machine prove its qualities better than any words possibly could. It has been ridden a mile in 2 min. 30 sec.

SPECIFICATION.

WHEELS.—Front made to suit reach ; back, 18 in. ; special racing spokes and patent hollow rims ; $\frac{9}{16}$ in. and $\frac{1}{2}$ in. best grey racing rubbers.

BEARINGS.—Improved double row balls to front wheel, and balls to back wheel ; also patent ball bearings to steering head and pedals.

FRAMEWORK, &c.—Of best weldless steel tube throughout ; handle bar bent down $5\frac{1}{2}$ in., best possible position for racing ; cranks adjustable from $4\frac{1}{4}$ in. to $5\frac{3}{4}$ in. throw.

FINISH.—Beautifully enamelled plain black ; handle bar, cranks, pedals, &c., highly nickel plated ; sent out with racing ball pedals, saddle, oil can, and necessary wrenches.

PRICE, all sizes, including Ball Pedals, £18.

London Depot: 32, Holborn Viaduct, E.C.

road, and to concentrate on maintaining that fast, relaxed and perfect rotary motion of the legs which makes for sustained speed. Everything else is poise. The body is balanced and breathing fully, the mind is fixed on 'stillness at the centre', the eyes are alert on the road as it whirls towards you. Everything moves. Everything is still. The challenge for all of us is to sustain this state for longer and longer moments in contact with everything that flows.

We seem to have come a long way from J. K. Starley's Rover, and yet the passion for lightness, mechanical excellence and silent speed has been a constant, even from the days of the high-wheeled Ordinary. The Rudge racing Ordinary of 1884 was as light as many modern competition machines (21.5 pounds) and it must have been quite an experience to travel at speed above a wheel nearly five feet high – 'an enjoyment which was *positively intense*', indeed. The 'experienced velocipedist' got it right with those words, for, in the end, the bicycle is as much a state of being as an object. In its design and its materials it represents a meeting of conservatism and innovation. In its construction it marries the skills of handcraft and precision engineering. Its philosophical principle contains the rotary perfection of the circle and the pursuit of 'lessness' – that fascinating concept of creating and controlling invisible forces with the minimum of material aid. Its integrity, its silence and, above all, its consistently and fundamentally human scale, make it a rare creation indeed.

But perhaps the balance needs to be redressed after so much idealism, and it is the Irish writer Flann O'Brien who gives us timely warning of a more sinister side to the bicycle and its intimate relationship with us. In *The Third Policeman,* Sergeant Pluck is explaining why he protects his parishioners by stealing their bicycles from them:

'Did you ever discover or hear tell of the Atomic Theory?' he enquired.

'No,' I answered.

He leaned his mouth confidentially over to my ear.

'Would it surprise you to be told,' he said darkly, 'that the Atomic Theory is at work in this parish?'

'It would indeed.'

'It is doing untold destruction,' he continued, 'the half of the people are suffering from it, it is worse than the smallpox.' . . .

'The Atomic Theory,' I sallied, 'is a thing that is not clear to me at all.'

'Michael Gilhaney,' said the Sergeant, 'is an example of a man that is nearly banjaxed from the principle of the Atomic Theory. Would it astonish you to hear that he is nearly half a bicycle?'

'It would surprise me unconditionally,' I said . . .

'Do you happen to know what takes place when you strike a bar of iron with a good coal hammer or with a blunt instrument?'

'What?'

'When the wallop falls, the atoms are bashed away down to the bottom of the bar and compressed and crowded there like eggs under a good clucker. After a while in the course of time they swim around and get back at last to where they were. But if you keep hitting the bar long enough and hard enough they do not get a chance to do this and what happens then?'

'That is a hard question.'

'Ask a blacksmith for the true answer and he will tell you that the bar will dissipate itself away by degrees if you persevere with the hard wallops. Some of the atoms of the bar will go into the hammer and the other half into the table or the stone or the particular article that is underneath the bottom of the bar.'

'That is well-known,' I agreed.

'The gross and net result of it is that people who spent most of their natural lives riding iron bicycles over the rocky roadsteads of this parish get their personalities mixed up with the personalities of their bicycle as a result of the interchanging of the atoms of each of them and you would be surprised at the number of people in these parts who nearly are half people and half bicycles.'

I let go a gasp of astonishment that made a sound in the air like a bad puncture.

'And you would be flabbergasted at the number of bicycles that are half-human almost half-man, half-partaking of humanity.' . . .

'And what way do these people's bicycles behave?'

'These people's bicycles?'

'I mean these bicycles' people or whatever is the proper name for them – the ones that have two wheels under them and a handlebars.'

'The behaviour of a bicycle that has a high content of humanity,' he said, 'is very cunning and entirely remarkable. You never see them moving by themselves but you meet them in the least accountable places unexpectedly. Did you never see a bicycle leaning against the dresser of a warm kitchen when it is pouring outside?'

'I did.'

'Not very far away from the fire?'

'Yes.'

'Near enough to the family to hear the conversation?'

'Yes.'

'Not a thousand miles from where they keep the eatables?'

'I did notice that. You do not mean to say that these bicycles *eat food*?'

'They were never seen doing it, nobody ever caught them with a mouthful of steak. All I know is that the food disappears.'

'What!'

'It is not the first time I have noticed crumbs at the front wheels of some of these gentlemen.' . . .

'How would you know a man has a lot of bicycle in his veins?'

'If his number is over Fifty you can tell it unmistakable from his walk. He will walk smartly always and never sit down and he will lean against the wall with his elbow out and stay like that all night in his kitchen instead of going to bed. If he walks too slowly or stops in the middle of the road he will fall down in a heap and will have to be lifted and set in motion again by some extraneous party. This is the unfortunate state that the postman has cycled himself into, and I do not think he will ever cycle himself out of it.'

'I do not think I will ever ride a bicycle,' I said.

4 Some More History, Technical and Social

First, a good bicycle had to be invented or evolved. Then there had to be factories capable of making the machine in quantity, and a large group of people with money, leisure and no social inhibitions about pedalling a less-than-thoroughbred iron horse. In Europe and America all these factors came together no less than twice in the nineteenth century: in the velocipede craze of 1868 and in the Safety Bicycle boom of the nineties. But they depended on what had gone before.

'The greatest invention of the nineteenth century was the invention of the method of invention,' according to the philosopher A.N. Whitehead. Invention could lead to a pot of gold and there were plenty of men looking for that. Elias Howe, a Massachusetts instrument-maker, one night dreamed he was being chased by a savage waving a spear with a hole in its point. His nightmare led to the first practicable sewing machine. A decade or so later, in 1860, 111,000 sewing machines were produced in America, nearly equalling in value all the other textile machinery made in that year. This incredible expansion led to a whole new power-tool industry, capable of making identical small precision parts. It drew heavily on the machinery and the techniques of mass production already devised in the small-arms and clock industries, just as power tools themselves had come from the railway boom of the 1840s. After the professional garment-makers had all been supplied, an aggressive force of 3,000 salesmen set about selling machines to the American housewife. It was from this relatively sophisticated phase of industrialization and marketing that the manufacture and true success of the bicycle arose.

There are many claimants to the honour of having invented the bike, and its history goes back some time before this industrialization. It is a history characterized by steady and slow agglomeration of ideas and innovations, including many false starts and technical cul-de-sacs.

There had always been dreamers. One of Leonardo da Vinci's unpublished drawings seems to include a bicycle-like invention. Another supposed early intimation is a trumpet-blowing cherub seated on a wheeled bench, a detail in a seventeenth-century stained-glass window in Stoke Poges Church near Windsor. Art historians see other than bicycle significance in this, though it is probable that the cherub is playing with a hobby horse, a common enough child's toy from the Middle Ages onwards. The more practical precursors of the velocipede, and indeed many of the velocipedes themselves, were the work of blacksmiths and wheelwrights, either riding their own inventions or constructing machines designed by ingenious patrons. Let us look at this 'pre-history' a little more closely.

BLACKSMITHS AND THE DANDY HORSE

Certainly the first 'bicycles' were, in effect, adult toys. An odd little craze was initiated in 1791 by a Monsieur (or Comte) de Sivrac, who put a second wheel and a padded saddle on the hobby horse and started a fashion amongst rich young Parisians for scooting and striding these cumbersome machines around the gardens of the Palais Royal and racing them along the Champs Elysées. This *vélocifère* or *célérifère* was vitally improved in 1817 by Baron von Drais de Sauerbrun of Mannheim, who made the front wheel steerable by the handlebars rather than fixed rigidly in the frame. He took his machine seriously, using it for the tours of inspection which he had to make as Master of the Woods and Forests of the Duke of Baden, and demonstrated it in Paris and London. A Mr Clarkson took out a patent for such a machine in America. Thus began the first international cycling boom, only a small one and the subject of much humour, scorn and satire. Nonetheless it established the idea

Hobby horse or dandy horse, c. 1820. It has large wheels and a steerable front fork, and the frame can be adjusted for height. They were usually made by wheelwrights and carpenters. This one is rather elegant.

of a two-wheeled self-propelled vehicle and created a foundation on which ingenious inventors and engineers could build and innovate if they so wished. The really important developments did not occur for half a century or so.

Riding the Draisienne, or dandy horse, was a craze of some substance, though chiefly restricted to the rich, young and idle if the caricaturists of the day, Rowlandson and Cruikshank for example, are to be trusted as impartial social observers. The machines were expensive enough if purchased new. In London Denis Johnson, a coach-builder, sold his version of the Draisienne, which he called a 'pedestrian curricle', for between eight and ten guineas, and he set up a riding school in Soho. Cartoonists took the idea seriously enough to suggest the anger of blacksmiths and horse-dealers at anything that might destroy their trade, though this may have been a fanciful possibility rather than a fact. Races were run. A four-horse coach was beaten by half an hour on the trip to Brighton, an alarming feat considering that the dandy horse weighed anything from fifty pounds upwards, though of course all road transport was

Two miles from London on what John Keats called the 'nothing' of the day. Several machines have elbow-rests to thrust against and there seems to be a tandem in the background. A three-wheeler with a passenger seat, if there ever was one, must have been very slow.

slowed by the road surfaces. Nevertheless some rural postmen seem to have been supplied with these vehicles in England. Baron von Drais fitted the first cycle accessory, a mileometer of sorts, to his machine in 1825. As late as 1832, after the craze had more or less died away, the Baron was still trying to revive interest in his invention during a visit to England. But for people with foresight in the 1830s the railways were an infinitely more exciting development, and indeed it must have seemed unlikely that future transport would use roads at all when rails were clearly more convenient and efficient.

As well as de Sivrac and von Drais, there are several other claimants to the honour of inventing the bicycle. A Frenchman, Joseph-Nicéphore Niepce, a pioneer photographer, claimed to have built a steerable *célérifère* before von Drais. More radically, the Russians date the first two-wheeled velocipede with pedals as early

Gavin Dalzell, a cooper from Lanarkshire, made this copy of Macmillan's design in 1845. The transmission seems to be out of line in this photograph. Fifteen years later Thomas McCall of Kilmarnock was producing similar machines for sale at £7 each. One of them is now in the Science Museum, London.

as 1801, the work of a serf and master craftsman named E. M. Artamonov. Perhaps the best-documented claim is that the first true bicycle emerged in Lowland Scotland about 1840, the creation of Kirkpatrick Macmillan, a blacksmith of Courthill near Dumfries. Only copies of his machine exist, which may be improvements of the original model, but the general shape of his machine, its originality and its ingenuity are without doubt. It was steerable, it had almost equally sized wheels and the rear axle was turned by treadles. Macmillan also broke new ground in being fined five shillings for riding on the Glasgow pavements and knocking over a child. These prelusive glimpses of the modern bicycle are fascinating, but they are chimerical so far as the general drift of bicycle history is concerned.

Along with the steady growth of industrialization, the last century also witnessed the gradual shift from amateurism in the sciences towards professionalism and specialization. A significant element in the development of the bicycle was the technological zeal of the industrial artisans, men with the same kind of curiosity, ingenuity and practical engineering skill as Macmillan. Early in the century such men had founded the so-called 'steam-intellect societies', along with the Mechanics Institutes and periodicals for disseminating new techniques and discoveries, such as the *Mechanic's Magazine* or, later, the *English Mechanic*. It seems to have been largely this section of the ingenious public which worried at the idea of self-propelled road vehicles, and which designed, constructed and tested them.

During the middle part of the century the mainstream of this kind of invention was concerned with three- and four-wheeled vehicles rather than the two-wheeler. A tradition of man-operated carts and carriages goes far back into the sixteenth century, if not further. Such novelties had been the playthings of royal families all over Europe at different times, each one uniquely designed and mostly intended to display the cleverness of the inventor. In contrast the Victorian experimenters saw their velocipedes as genuine means of travel or recreation, although to begin with their machines were heavy, complicated and often too large. Many can never have been operational. However, some successful builders went into business and made their machines for a clientele of enthusiasts, mostly rich. Probably the foremost of these small manufacturers was a Dover carpenter, Willard Sawyer, who included the Emperor of Russia amongst his patrons as well as French and German royalty, and who exhibited along with other firms making velocipedes at the Great Exhibition of 1851. Prince Albert preferred a machine built by J. Ward, a coach-builder in Leicester Square. Sawyer's catalogues include sociables, ladies' carriages, and lightweights for racing and touring. At prices ranging from £10 to £40 they were not cheap.

THE VELOCIPEDE

The bicycle seemed stuck at the dandy horse, apart from a few unique specimens, until a Parisian manufacturer of three-wheeled velocipedes, invalid carriages and perambulators took the machine a vital step forward. Pierre Michaux, with the help of his son Ernest, attached cranks and pedals 'like a grindstone' to the front wheel of an old Draisienne. Thus they too can claim to have invented the bicycle. Their idea was not in fact entirely original, but Michaux had the luck or insight to market his machine at a ripe moment and having found a receptive public he had the business acumen to expand fast. The experiments were made in 1861. In 1863 his company made 142 velocipedes. After the Paris Exhibition of 1867 the demand grew hugely, and by the next year Michaux had built a new factory employing about 300 men who completed five velocipedes a day.

By that time many other firms were also making money out of the boom. For money was surely to be made from a craze such as this, a fact that helps to explain the epidemic spread of interest in the new machines through France to Britain and America. By 1868 an American firm was exporting its Pickering Improved Velocipede to Liverpool, anticipating Rowley Turner and the Coventry Machinists' Company by a year or so. Inventors flooded into the patent offices with new ideas and the periodicals, especially the *English Mechanic*, which started in 1865, kept up an impatient interest in the technical development of the new machine. In 1869 the magazine actually published drawings of a design by F. W. Shearing for a bicycle with a chain-driven rear wheel. It was never built, however.

Every innovator hoped to build a fortune on the strength of his ingenuity. An assistant of Pierre Michaux's, Pierre Lallement, began an acrimonious dispute with his erstwhile employer about who should take credit for the invention of the velo. Lallement had rushed to America, taken out the American patent for the new machine, failed to market it successfully, returned to France when he saw the boom there was well under way and established a firm to rival the Michaux concern. But he seemed doomed not to make money in the long term. A New York velocipede-maker, Calvin Whitty, cunningly purchased the patents from Lallement for $2,000 and then proceeded to collect a royalty of $10 on every machine made in the USA until the collapse of the boom.

Other inventors went unrecognized. André Guilmet, a clock-maker killed in the Franco-Prussian War, seems to have left behind him an improved 'boneshaker' with chain drive, wire spokes and rubber tyres. But like so many early designs it was a prototype only and the original cannot be substantiated. Guilmet is only one of several 'begetters' of the modern two-wheeler whose claims are advanced according to the bias and the nationality of the historian. Michaux himself lost his fortune and died in ignominy. He had sold his company to his partners and retired early. But he started

In 1869 the first ever road race was ridden on machines like this over the 76 miles between Paris and Rouen. Englishman James Moore won, out of 100 starters. This velocipede was made by K. W. Hedges.

manufacturing again in his own name, was successfully sued for breach of contract and died penniless in a Paris hospital for the poor and insane.

The Coventry Machinists' Company, which started full production of velocipedes in 1869, originally intended exporting its machines to a safe market in France, but within a year the Franco-Prussian War put an end to this project. Rowley Turner, who was the Company's French agent, is said to have escaped from Paris on one of his own firm's products. In Britain he found an equally responsive market for the new boneshakers. Like Michaux, Turner judged the moment correctly and went into production.

War temporarily halted the French industry, which turned to making armaments. Consequently the British industry took the lead in technical development, an initiative which it kept until the bicycle had arrived at more or less its modern form. In the same way, forty years later, the Great War was to give America the initiative in the production of the motorcar.

The boom in velocipeding lasted only a few years. Interest in two-wheeled transport remained, especially in Britain, but cycling soon ceased to be of much news interest, except to an established band of devotees for whom it was a leisure occupation. Velocipedomania had been the rehearsal for the real bicycling explosion which was still to come. Like the dandy horse, the boneshaker was first taken up as a diversion by the relatively well-to-do, and there was more wealth generally for pastimes of this sort than in the Regency period. The careers of the velocipede manufacturers themselves testify to the new class of entrepreneur, not necessarily hugely wealthy, that had been created by the persistent increase in industrial activity during the nineteenth century. As second-hand machines became available, the less well-off took their turn on the boneshaker. Old models were soon superseded by faster and more technically advanced mounts. A general pattern of manufacture and of recreational and sporting use for the bicycle had been established.

THE ORDINARY

What were the forms through which the bicycle still had to evolve? In the search for both speed and comfort the rear wheel shrank and the front wheel grew larger and larger until it reached the greatest possible diameter for direct drive to the hub, dictated by the leg measurements of the rider. Five-foot-high wheels could not be swallowed up by even the largest pot-holes, and at every pedal stroke the rider of the new high-wheeler travelled a good deal further than on any old-fashioned boneshaker. The Ordinary, or Penny Farthing, as it was later nicknamed, was the steed of the young and athletic male cyclist from the early 1870s to about 1885. Technical ingenuity, understandably enough, concentrated attention on the construction of the wheels. Ball bearings became common. Instead of the stiff wooden radial spokes of the velocipedes, the hub was suspended in the wheel by wire spokes. Systems of tensioning and attachment varied until the tangent method still used today was invented by James Starley in 1876. Solid rubber tyres were stuck or sprung onto the rims. The Ordinary was elegant, light and racy. To begin with it was as expensive as the boneshaker at about £20, but gradually throughout the period of its use it became more and more possible to find cheap new machines, even for less

'Elegant, light and racy', the Howe Spider was built in Glasgow, c. 1885. Note the mounting step, the sprung saddle and the hollow oval tubes of the frame and forks. The lever on the handlebars operates a typical spoon brake on the front tyre.

than £5, and hire purchase put these within the reach of a new kind of rider.

The Ordinary was also dangerous. Mounting was achieved by scaling a step at the back of the machine while scooting with one foot. (The velocipede itself, with large heavy wheels and fixed cog, must also have been an unwieldy beast on which to push off.) Controlled stops on an Ordinary necessitated springing backwards off the machine and grabbing it from behind. Those who rode it would at least have been accustomed to the idea, if not the management, of a horse, on which the rider is also elevated far from the

The Singer Xtraordinary of 1878. The lever cranks allow the rider to sit farther back on a large wheel, thus lessening his chances of taking a header. The sloping front forks improve his directional stability even more. These machines were finely made with forged steel axles, bronze hub flanges and adjustable ball bearings, but the 'elbows' on the cranks sometimes turned themselves inside out, and the finished product was rather heavy.

ground and in a situation which he cannot always predict with certainty. The Ordinary's problem was the 'header' or 'imperial crowner'. The rider sat almost vertically above the hub in order to get all his weight and strength on the pedals; this, plus the high centre of gravity, meant that if the wheel met any serious impediment the rider took off over the top. One recommended method of descending hills at speed was to hook the legs over the top of the handlebars, so that in the event of a header the rider would be thrown neatly clear of his machine.

In spite of these apparent disadvantages the Ordinary had its devotees, and it was used extensively for touring and racing. Clubmen from the towns explored the country lanes dressed in paramilitary uniforms. They rode together not just for the pleasure of companionship, but because the new phenomenon of cycling had its natural enemies. Perched as they were vulnerably far from the ground, they were subject to arbitrary attacks from the general public. The most celebrated case in Britain concerned the guard of the St Albans mail coach, aptly named Henry Cracknell, who devised a kind of bicycle-bolas, an iron ball on the end of a rope, which he could hurl into passing wheels. He was brought to court after successfully demolishing the Trafalgar Bicycle Club and fined a mere £5, but even this was a moral victory for the beleaguered cycling minority.

Horses were frightened by bicycles and cyclists were therefore unpopular with horsemen, and especially with the professional wagon-drivers and teamsters, who felt, with some justification that they ought to own the road, and in practice did. Accustomed as we are to fixed rules of driving on one side of the road, assisted by every conceivable kind of signpost and road furniture, it is difficult to imagine the mêlée that could clog city centres before the orderly traffic jams of the mid twentieth century. Cyclists stood little chance against the four- or six-horse goods wagons, whose drivers liked to stop, start and turn just whenever and wherever they chose. Cabmen were similarly unsympathetic, not least because they saw the two-wheeler as a threat to their trade. As pavements were often the smoothest and safest places to ride, pedestrians became yet another natural enemy for the cyclist. Even when he did find the freedom of the open road he might be stoned by urchins, chased by dogs and eventually turned away from places of refreshment and hotels. The national Bicyclists' Touring Club was formed in 1878 to protect the rights of the cyclist and to disseminate information about roads, hotels and events in the cycling world.

Whereas attacks on passing cyclists by mailcoach drivers or irate huntsmen might be sensationalized in the papers, the curious legal status of the bicycle led to police harassment which was even more persistent and irritating. From its advent no one had been quite sure whether the two-wheeler was a vehicle or not, from a legal and administrative point of view. In Britain public hostility was focused after the passing of the Highways and Locomotives (Amendment) Act, the first British legislation to refer specifically to the bicycle. This handed power to county authorities to regulate the use of cycles, and the resulting variety of by-laws was bewildering to all and often used unfairly.

1880. *Bicycling Notes – No. 1*. 'The police have strict orders to arrest any Bicyclist riding without a bell or whistle.'

The CTC campaigned vociferously to end this muddling state of affairs and eventually, in 1888, cycling MPs successfully added a clause (Section 85) to the Local Government Act, which did away with the county authorities' powers, repealed all local laws relating to cycling and declared all bicycles, tricycles and velocipedes 'to be carriages within the meaning of the Highway Acts'. This legislation, nicknamed in the *Law Journal* the *Magna Carta de Bicyclis*, meant the cycle had legally come of age, and the cycling lobby was consequently made more powerful. But this is to anticipate.

When, in 1883, the BTC had been renamed the Cyclists' Touring Club, the change was not just a matter of whim. 'Bicyclists' meant riders of the Ordinary, for during its hegemony this was the only kind of bicycle available. It came to be called Ordinary later only to distinguish it from the Safety. Because of the exclusive character of their mount, the bicyclists had tended to become a group apart from the other cyclists using the road. The less athletic and the more respectable enthusiasts of self-propulsion rode tricycles or quadricycles, machines which, keeping their wheels three- or four-square on the road, were not so liable to spectacular prangs, and which gentlemen and even ladies could mount without

'A carriage within the meaning of the Highway Acts.'

undignified leaps and bounds. After several abortive experiments with side-saddle bicycles for the ladies' market, it had been the inventive James Starley of the Coventry Machinists' Company who came up with the Coventry Lever Tricycle, the first three-wheeler design successful enough to go into large-scale production. Starley added an extra wheel to his basic model to create sociables and quadricycles. One of his final tricycle designs, the Salvo, received the ultimate accolade when Queen Victoria, seeing one in action near Osborne, demanded first to meet the young lady who owned it and then Starley himself, who was asked to deliver such a machine in person to her majesty. Thenceforth he called it the Royal Salvo.

James Starley in 1887 on his Salvo tricycle, the first to be fitted with his 'balance gear' and a very popular and influential design. Properly termed a 'quadricycle', it is steered by the rear wheel while the little wheel at the front prevents capsizing. Later models steered at the front. Note the cranks and the chain drive.

With the design of the Ordinary more or less static for a few years at least, cycle engineers concerned themselves with the special problems of the multi-wheelers. The sociables employed a 'balance gear', refined and patented by Starley, to equalize the effort of the two pedallers. This later served as the basis for the differential on

An 1885 Rudge Rotary tricycle. The large wheel is driven by a chain and the machine is steered by the stirrup handle on the right which turns both the smaller wheels together. The design goes back to 1878 and it remained popular for ten years.

motorcars. Continuous chain drive was also a vital pioneering feature, one which was to point the way to the Safety. From the mid-eighties onwards tricycle design coalesced with the appearance of the new Safety machines, which were beginning to proliferate, and the present-day shape of both bicycles and tricycles was soon to be established.

Various hybrid or dwarf Ordinaries with geared-up wheels began to appear round about 1884. They had mildly apologetic names, like the Bantam or the Kangaroo, and they were met with derision.

Many a clubman refused to believe that these squat and graceless machines could be faster than his own majestic mode of locomotion. The bicycle companies employed racing cyclists to break records on the new machines and gradually the contempt changed to interest and eventually enthusiasm.

Kangaroo by Hillman, Herbert and Cooper, Coventry, 1884. Each front fork is extended below the axle to support a pedal and a crank. The wheel is geared up and driven by chainsets fitted on both sides.

THE SAFETY

The way was clear for numerous early versions of the Safety. In 1876 George Shergold, a shoemaker of Gloucester, built and rode a bicycle with pedals and a chain drive to the rear wheel. But he only made one of them. H. J. Lawson's 'Bicyclette' had had the same features as early as 1873 but it took seven years before the Tangent Bicycle Co. brought it into a limited production. Other firms were quicker off the mark and Coventry Machinists, along with Humber

The evolution of the Rover Safety. *From the top*: 1884: the cycle has a 36-inch front wheel and indirect steering. The 'top tube' is really a bracket for the saddle which can be adjusted for height by raising or lowering the whole tube, clamped at each end. 1885: Direct steering and sloping front forks. The saddle bracket/top tube (seen again here) was very soon replaced by an L-shaped seat-pin, and the cycle reached its final pattern as shown below. The Rover went through eight different models with increasingly strong frames and including curved front forks.

and the Birmingham Small Arms Company, led the way. John Kemp Starley, the nephew of James Starley, produced the first 'Rover' Safety in 1884, and a year later it featured a diamond frame. By 1890 this was all but the standard design, having proved stronger than the elegant cross frame used in Lawson's and other models.

One vital improvement still had to be made to improve the comfort of the rider, and for once it was not Starley who made the necessary invention. This was the pneumatic tyre, first patented in 1845 by a 23-year-old Scottish engineer called Robert W. Thomson. The young man invented his air-filled and solid tyres for horse-drawn wagons, railway rolling stock and traction engines. He was too far ahead of his time, however, and could not raise enough interest to make a commercial success. It took the advent of the Safety bicycle and the motorcar to vindicate his foresight.

In 1887, fourteen years after Thomson's death, the idea for a pneumatic tyre for bicycles occurred independently to John Boyd Dunlop, another Scot, working as a veterinary surgeon in Belfast, and

H. J. Lawson's Bicyclette, 1879. The patent mentioned the possibility of direct steering from the top of the sloping front forks, but it was Starley's Rover which actually realized this feature in the Safety.

he fitted a prototype to his young son's machine. Within two years William Harvey du Cros had founded the Pneumatic Tyre Company to produce Dunlop's idea. These were hectic times with many problems over patent rights and dozens of different designs and claims from rival companies. The early tyres had been bound and glued more or less permanently to the wheel rims and it was not until the invention of an easily detachable cover held on by two hoops of wire at the edges that 'Dunlops' truly came to command the marketplace.

Like the cycle manufacturers du Cros saw the value of publicity through sport, and his company fielded a racing team on pneumatic-tyred Safeties from the very start. So successful were these machines against the racing Ordinaries of the day that the founder's son Arthur, the Champion of Ireland, was once banned from competition in England. Pneumatic tyres were fitted to some high-wheelers too, but the problems were, literally, enormous. Of course race tracks had better surfaces than the public highway, and punctures were an all-too-frequent occurrence for the everyday rider. Clubmen scoffed loudly at the fragile and unaesthetically fat 'sausage'

Singer cross-frame bicycle, 1888. Like most fixed-wheel models of the period, it has pegs on the front forks to rest the feet while going down hill. The long sloping forks make for stability.

tyres, comparing them unfavourably with their own slim wheels. Despite all their jeers the combination of safety frame and air-filled tyre was soon to be unbeatable.

THE BICYCLE BOOM

For the last decade of the nineteenth century the bicycle ruled the road in the modern world; for the first time in the history of the world there existed a thoroughly convenient mode of transport available to all classes of society, used alike by men and women, young and old.

Flora Thompson, in her reminiscences of village life at the turn of the century, *Lark Rise to Candleford*, records both the strangeness and relative shortness of the Ordinary era and the effects of the boom in cycling when the Safety arrived. Her descriptions are evocative and worth quoting at length:

But, although it was not yet realised, the revolution in transport

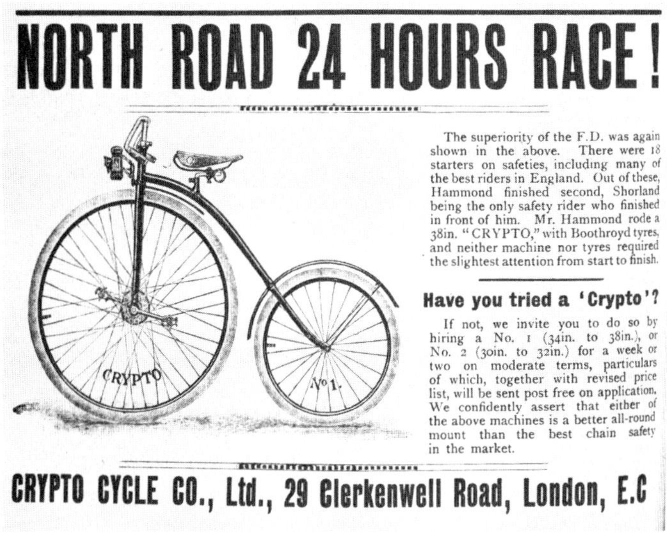

Front-wheel drive died hard and Crypto cycles equipped with pneumatic tyres were advertising racing victories as late as the mid 1890s.

had begun. The first high 'penny-farthing' bicycles were already on the roads, darting and swerving like swallows heralding the summer of the buses and cars and motor cycles which were soon to transform country life. But how fast those new bicycles travelled and how dangerous they looked! Pedestrians backed almost into the hedges when they met one of them, for was there not almost every week in the Sunday newspaper the story of someone being knocked down and killed by a bicycle, and letters from readers saying cyclists ought not to be allowed to use the roads, which, as everybody knew, were provided for people to walk on or drive behind horses. 'Bicyclists ought to have roads to themselves, like railway trains' was the general opinion.

Yet it was thrilling to see a man hurtling through space on one high wheel, with another tiny wheel wobbling helplessly behind. You wondered how they managed to keep their balance. No wonder they wore an anxious air. 'Bicyclists' face', the expression was called, and the newspapers foretold a hunch-backed and tortured-faced future generation as a result of the pastime.

Cycling was looked upon as a passing craze and the cyclists in their tight navy knickerbocker suits and pillbox caps with the badge of their club in front were regarded as figures of fun. None of those in the hamlet who rushed out to their gates to see one pass, half hoping for and half fearing a spill, would have believed, if they had been told, that in a few years there would be at least one bicycle in every one of their houses, that the men would ride to work on them and the younger women, when their housework was done, would lightly mount 'the old bike' and pedal away to the market town to see the shops. They would have been still more incredulous had they been told that many of them would live to see every child of school age in the hamlet provided by a kind County Council with a bicycle on which they would ride to school, 'all free, gratis, and for nothing', as they would have said.

Later in the book she expands her comments on the cycling clubs and the not-so-gradual involvement of women in the pastime:

The sound of a bicycle being propped against the wall outside was less frequent than that of a horse's hoofs; but there were already a few cyclists, and the number of these increased rapidly when the new low safety bicycle superseded the old penny-farthing type. Then, sometimes, on a Saturday afternoon, the call of a bugle would be heard, followed by the scuffling of dismounting feet, and a stream of laughing, jostling young men would press into the tiny office to send facetious telegrams. These members of the earliest cycling clubs had a great sense of their own importance, and dressed up to their part in a uniform composed of a tight navy knickerbocker suit with red or yellow braided coat and a small navy pillbox cap embroidered with their club badge. The leader carried a bugle suspended on a coloured cord from his shoulder. Cycling was considered such a dangerous pastime that they telegraphed home news of their safe arrival at the farthest point in their journey. Or perhaps they sent the telegrams to prove how far they really had travelled, for a cyclist's word as to his day's mileage then ranked with an angler's account of his catch . . .

They were townsmen out for a lark, and, after partaking of refreshment at the hotel, they would play leap-frog or kick an old tin about the green . . . Soon, every man, youth and boy whose families were above the poverty line was riding a bicycle. For some obscure reason, the male sex tried hard to keep the privilege of bicycle riding to themselves. If a man saw or heard of a woman riding he was horrified. 'Unwomanly. Most unwomanly! God knows what the world's coming to,' he would say; but, excepting the fat and elderly and the sour and envious, the women suspended judgement. They saw possibilities which they were soon to seize. The wife of a doctor in Candleford town was the first woman cyclist in that district. 'I should like to tear her off that thing and smack her pretty little backside,' said one old man grinding his teeth with fury. One of more gentle character sighed and said: "T'ood break my heart if I saw my wife on one of they,' which those acquainted with the figure of his middle-aged wife thought reasonable.

Their protestations were unavailing: one woman after another appeared riding a glittering new bicycle. In long skirts, it is true, but with most of their petticoats left in the bedroom behind them. Even those women who as yet did not cycle gained something in the freedom of movement, for the two or three bulky petticoats formerly worn were replaced by neat serge knickers – heavy and cumbersome knickers, compared with those of today, with many buttons and stiff buttonholes and cambric linings to be sewn in on Saturday nights, but a great improvement on the petticoats.

And oh! the joy of the new means of progression. To cleave the air as though on wings, defying time and space by putting what had been a day's journey on foot behind one in a couple of hours! Of passing garrulous acquaintances who had formerly held one in one-sided conversation by the roadside for an hour, with a light *ting*, *ting* of the bell and a casual wave of recognition.

At first only comparatively well-to-do women rode bicycles; but soon almost everyone under forty was awheel, for those who could not afford to buy a bicycle could hire one for sixpence an hour. The men's shocked criticism petered out before the *fait accompli*, and they contented themselves with such mild thrusts as:

Mother's out upon her bike, enjoying of the fun
Sister and her beau have gone to take a little run.
The housemaid and the cook are both a-riding on their wheels;
And Daddy's in the kitchen a-cooking of the meals.

And very good for Daddy it was. He had had all the fun hitherto; now it was his wife's and daughter's turn. The knell of the selfish, much-waited-upon, old-fashioned father of the family was sounded by the bicycle bell.

With the wholesale manufacture of the Safety the bicycle could become a means of transport rather than a leisure or sporting novelty. Until this time those who rode bicycles were defined in the first phase by their mechanical ingenuity and afterwards, in later phases, by their desire for speed and their athletic ability. For the first few years of the cycling boom the bike's predominant role was still as an instrument of leisure. As it became less of a rarity its use grew more utilitarian. But in 1895 and for the two years following, cycling was a craze not just amongst the urban middle class, but with the aristocracy and with proponents of fashion and high society. In London the wealthy pedalled through the Parks with crowds thronging the walks to watch the stately spectacle. In Paris the Bois de Boulogne served the same purpose. The fashion papers avidly discussed the most suitable costume for the new activity. The horrid spectre of rational dress, with terrible tales from America of bloomers, and perhaps worse, hovered in the background. Lady novelists decried the habit as vulgar. Actresses posed by their machines. In America a few incredibly costly bicycles with gold and silver plating and encrusted with jewels were bestowed by millionaires on their lady-friends. High society adopted the bicycle with fervour and made it news – perhaps only for a couple of years or so, yet even this short span was a considerable length of time in comparison with today's fads. Historians of cycling are apt to disparage Society's flirtation with pedal-power, but, by giving the bike much-needed prestige and style, it eased and quickened the growth in popularity of the activity.

In Britain the bicycle had its detractors but there seems not to have been quite the same faddish worry at the possible ill-effects that filled the newspapers in America. Up till 1893 doctors had maintained that cycling was a notorious cause of illness. The medical profession worried deeply about 'bicycle walk', 'bicycle heart' or, most dreaded of all, *kyphosis bicyclistarum*, 'cyclist's hump', caused by strenuous pulling on the handlebars. The bike was also blamed for unemployment amongst tailors and hatmakers, decline in piano-playing, increase in sabbath-breaking and non-attendance in church. The church was in doubt about its morality. But some churchmen, no doubt wanting to keep on good terms with their congregations, supported the new pastime. In New York City a Presbyterian argued that agnosticism often had its origins in dyspepsia, for which cycling was a perfect cure, while in Brooklyn a Baptist went even further, calling the bicycle a 'scientific angel which seems to bear you on its unwearied pinions'. For him the bicycle was a symbol of man's perfection and regeneration through Christ.

In Europe and America industry worked overtime to supply the

High society and fashion in Battersea Park, 1895. The lady dismounted on the right is wearing a daringly masculine 'rational' outfit as pioneered two years earlier by the 'scandalous' Miss Reynolds, the 'Brighton Female Scorcher'.

A WARNING TO ENTHUSIASTS.

Punch, 1889

boom. Over-capitalization and over-production to keep pace with a demand that levelled out rather than increased led to a vast slump in 1897, with many bankruptcies and firms going out of business. The biggest firms, such as that of Albert A. Pope in America, one of the earliest and largest mass producers, started the process of buying up others and aimed for monopoly. The subsequent price wars radically cut the cost of the bicycle in America from over $100 to $30, although somewhat less dramatically in Britain. The result was that cycling thrived and the bike came within the purchasing power of yet more people. Membership of the CTC provides a handy indication of the general trends. The advent of the Safety doubled the membership, from 16,343 in 1895 to 34,655 in 1896. From then on the figures crept up steadily until 1899, when the CTC had its highest total of members ever, 60,449. This number represented only the enthusiastic cycle tourists, the tip of an iceberg so far as overall numbers are concerned.

By 1900 the craze for cycling was over: it was now no longer a novelty but a fixture of ordinary life. Yet even in that year the Bureau of Census in the USA reported that 'few articles ever used by man have created so great a revolution in social conditions as the bicycle.'

SOCIAL CHANGE AND THE 'WOMAN QUESTION'

What was this revolution that the bike had effected? It would be delightful to write of these five years as a kind of Utopian island in time when workers and aristocrats, men and women mounted for the first time in history on thoroughly democratic machines, participated in fearless and cheerful social exchanges about sprocket sizes and the dangers of steep hills as they pedalled healthily through the leafy highways and byways until their peaceable kingdom was shattered by the motorcar, that dust-raising, polluting and socially divisive mechanism. Flora Thompson's emphases are right, however, and they are supported by the facts. The boom saw townsmen invading the countryside as a leisure pastime. It saw the social classes taking a spin on the bicycle each in turn, more or less in pecking order, though preceded by enthusiasts of no particular class. It saw a real and important break with the past in the transport emancipation of women. And it saw the beginnings of a world where transport would be available for nearly everyone, with the bike eventually left for the humble, after it had paved the way for the motorized machinery which would soon dominate the roads.

If literature may be used once more to evoke history, H. G. Wells's *The Wheels of Chance* is the document that most poignantly describes the pains and pleasures of the new pastime. Hoopdriver is certainly liberated by his machine. It affords him not only a country holiday, in itself a remarkable event which he enjoys immensely, however ignorant of the countryside he may be, but also a brush with a society girl, riding on pneumatics and wearing some kind of Rational Dress. The book suggests the new social mobility created by the bike, which breaks the boundaries of Hoopdriver's world both literally and figuratively. *The Wheels of Chance* is a flimsy romance with a mild twist of irony in that its hero never becomes Prince Charming. In novelette fashion the cad still menaces and the heroine weeps in innocent despair, while being saved from the full consequences of the plot. Realistically, the gulf between the classes is not bridged, except in the brief limbo created by the bicycle. But that limbo is still worth noting. Wells accurately characterizes the classes to whom the new freedom of movement meant most. Hoopdriver is a draper's assistant, genteel but far from being a gentleman, at the bottom end of the urban middle class, while Jessie is a bright if immature young woman suffering the privileged ache of modernity.

Hoopdriver has a second-hand machine of markedly old design, but it is his most valued possession and he estimates himself highly for having taken up the new leisure pastime. Here was no miracle leveller of the classes, however, and the draper's assistant is very conscious of better and newer models on the road. Nevertheless, the bicycle will not go without pedalling and so every cyclist is a 'worker' in this sense. It was the motor which brought back the special snobbery of the private carriage and the chauffeur. Thus Hoopdriver's metamorphosis into a 'bloomin' dook' the moment he masters his machine is not without cause, for 'scorchers' came from every class. At one time the Bishop of Chelmsford was the top sprinter in Essex.

The cycle companies made sure that their models underwent continuous modification and improvement, thus keeping price and prestige differences between the latest and last year's models. Cycle shows, yearly events in America and Europe during the boom, were later taken over by the car companies as an effective method of whipping up public interest in their new and ever slightly changing product. Hoopdriver's bike represents a big financial outlay. Early in the craze bicycles were expensive items, both new and second hand. Some American figures are suggestive. In 1893 new bicycles cost as much as $150, almost the same as a horse. This was approximately six months' pay for the factory hands who were making them. In 1897, just before the slump, the new bike averaged $80. Five years later they cost $3 to $15. In comparison the price of a horse, quite apart from the fact that the townsman lacked the space and time to keep one, rose continually throughout the nineties, and, once bought, still cost $150 a year to feed and groom, before it died.

Bicycles broke the oat barrier for ever. Horses, except in an agricultural context (Old Uncle Tom Cobley and all) had only really been available to those who had land and servants. Even by 1897 personal transport for all must have seemed a distinct possibility rather than a laughable dream or nightmare. This in itself was a revolution in social mobility. Indeed, in an average day's riding a man on a bicycle could travel almost twice as far as he could on horseback. Never before in the history of mankind had it seemed likely, or desirable, that commonplace, even relatively poor, individuals could cover large distances according to their own whims. This expectation, bicycle-created, has of course become the inheritance of the motorist in the modern Western world, though by no means over the whole globe. Ultimately, the bicycle's most important role was to create the idea of private rather than public transport for everyone, and in doing this it was the harbinger of a special kind of freedom.

An interesting insight into the role of the bicycle at this time is provided by the history of the cycle clubs. These had started as gentlemen's organizations. By the late 1890s the lower middle classes were forming their own clubs and, as the century turned, the Temperance Societies and the socialist Clarion Clubs had established cycling sections to cater for the new demand. A full history of these movements has yet to be written. As a relatively impersonal national association, the Cyclists' Touring Club probably remained a cut above the other clubs. The purpose and effects of the CTC's policies, its route maps, its hotel guides, and so on, were to tame the countryside and to provide the town-dweller with clearly marked roads where he might meet like-minded folk with whom to pass the time of day until he could put up his feet in a safe, clean and inexpensive hotel, a true home from home. These cyclists were amongst the first independent tourists in the modern sense, though of course the railways had led to the package tour and day trip many

The Biggar Cycling Club, c. 1890. The fork foot-rests on their cross-frame Safeties can be clearly seen. The gentleman fourth from the left is carrying the club bugle. Four Ordinaries stand in the back row.

years previously. (Thomas Cook ran his first organized excursion in 1845.) Characteristically, the CTC was host to the first annual congress of the Ligue Internationale des Associations Touristes in 1899.

In 1906 the CTC applied to the High Court to amend its constitution to admit all tourists, including motorists. Some of its officers had for some time been sniffing the exhaust fumes on the winds of change, and, reasonably enough, were seeking to widen the franchise of their club. The previous year a huge majority of the members (10,495 to 2,231) had voted to widen the club's basis. It may be inferred from this that many CTC members were in the vanguard of the new motorcar era, or at least hoped to be, and this at a time when cars were very expensive. The astute High Court judge who deemed that the club could not protect the interests of both cyclists and motorcar drivers at least understood that the hegemony of the cyclist was over. Whatever revolution had been begun by the bicycle was to be absorbed and overlaid by other far-

Horse-drawn and motor vehicles mingle in both private and public transport at Piccadilly Circus in 1910.

reaching changes, now that the plutocrats had focused their attention on the motorcar. The Touring Club de France, however, accepted the motorcar into its ranks.

Before the advent of the motorcar, back in the nineties, the bicycle was notorious in its own right as an emblem of changing attitudes, and not only towards transport and travel. An important factor in its notoriety, as suggested by Flora Thompson, was the issue of women and the bicycle. The problems of whether women should ride, what they should ride, how, in what clothes and with whom they should ride, occupied many a column of newsprint and, no doubt, not a few abusive bar-room conversations as well. Questions of etiquette suggest the new freedoms enjoyed by the lady cyclists. Should they accept help from unknown gentlemen (or even

MODERN ATHLETICS.—A DIPLOTRIBICYCLICAL QUARTET.
How this Drawing would have astonished Our Readers Twenty Years ago!

Punch, 1882

ordinary men) when broken down? Should ladies ride in front, and thus steer, on a tandem? Who went first down a hill? A 'Chaperone Cyclists' Association' was begun in 1896 to provide 'gentlewomen of good social position to conduct ladies on bicycle excursions and tours'. Chaperones had to provide their own bicycles and the Association was not responsible for accidents.

In the matter of clothing, agitators from the Rational Dress Society led the way. The bicycle became for them a vital focus for their arguments. The old styles not only hampered the female cyclist's movements, but could also be dangerous. Bloomers had been invented in the 1850s, some years before the advent of the Safety, and not in fact by Amelia Bloomer but by her friend Mrs Miller. They seem to have been a rarity in Britain and France, but in America they were commonplace. For the three years following 1895 they became high fashion for racy young women, gradually shortening and tightening during this time until they were almost like men's breeches. There was the usual brouhaha accompanying a risqué fashion. Restaurant owners made waitresses wear them. Actresses declared they would never sink so low as to be seen in them. Bloomer dances in Chicago were eventually banned when the

'. . . an emblem of changing attitudes'. In 1897 Cambridge undergraduates demonstrated against the proposal to admit women to degrees. As a symbol of what they feared, they hung a female effigy – complete with bicycle – opposite Senate House Passage.

police decided that anyone wearing the garment was a common prostitute.

Cyclists had almost always worn clothes that set them apart from the ordinary pedestrian and, in turn, the new recreation led to changes in fashion on the pavements. Thus for men the Norfolk

This lady with her Humber Roadster has not made many concessions to cycling dress; but the oil-cloth chaincase and the cords on the rear mudguard will protect her skirts.

jacket and breeches buttoned at the knee became respectable outdoor gear, supplanting the bulky frock coat. Top hats nearly became extinct. Clearly women needed some kind of cycling outfit as well, but could such clothes be decent for other occupations such as archery or tennis, let alone everyday wear? In Europe, perhaps more conscious of fashion and decorum than America, the convertible skirt (doubling as knickerbockers) and other genteel bifurcated garments seem to have been popular. Hoopdriver is astonished at the magical transformations Jessie can achieve with her dress, but does not begin to speculate on the mechanics of it.

The Rational Dress Society suffered much acrimony and many minor public defeats. In 1899 the CTC also got involved. Lady Harberton, the founder of the Rational Dress Society and a CTC member, had been refused admission to the coffee room of the Hautboy Hotel, Ockham, Surrey. The landlady, Mrs Sprague,

'A halt by the roadside for a quick smoke and slight repairs.' This remarkable photograph illustrates everything that the reactionaries must have disapproved of and feared. These girls are not upper-class 'ladies' and they look less like fashion plates and more like real cyclists.

taking exception to her ladyship's attire, had insultingly shown her to the bar parlour, a male preserve. The club's opinion was split about the justice of this, but most members felt that it was an outrage against cyclists, leaving the 'women question' out of the argument, and eventually Mrs Sprague was taken to court for unlawfully refusing to supply a traveller with victuals. In court Lady Harberton produced a photograph of herself in the objection-

Free movement at the turn of the century: the Safety bicycle, the pneumatic tyre, and comfortable, practical clothing for both sexes. No doubt Elliman's is a splendid embrocation for aches and bruises, but it is social 'stiffness' which has been banished.

able garments. It showed 'an elderly lady wearing a pair of exceedingly baggy knickerbockers reaching below the knee and a jacket which came well over the hips and opened sufficiently to reveal the silk blouse underneath'. Mrs Sprague, not to be outdone, provided a photograph of her bar parlour. She spoke of ladies in skin tights seen cycling on the Portsmouth road, and of her general decision to protect her custom by excluding rational dress from the coffee room. The jury quickly made up its mind that the bar parlour, as depicted in the photograph, was a perfectly suitable place for serving food and Mrs Sprague was acquitted, with the spectators

DIVISION OF LABOUR.

It is not the business of Ducal Footmen to Clean the Family Bicycles. The Ladies Ermyntrude and Adelgitha have to do it themselves.

Punch, 1896

applauding the decision. A public defeat for Lady Harberton and her cause, but time has given both her and Mrs Sprague moral victory: women have won both trousers and the public bar.

While Lady Harberton was making news, womenfolk in general moved steadily towards rationalization and away from the billowing yards of material and the deforming corsets worn by their grandmothers. The crinoline had been shed by 1870, and in 1890 the bustle went out of fashion just before the cycle craze. Despite all the gadgets and chain guards devised by manufacturers, and the lead-weighted skirt hems, ankles and calves did suffer some exposure; but after the first scandalized comment, no one seemed much worried. Society was obviously ready for a change in manners and morals so far as women were concerned and the bicycle fitted conveniently into a wider changing pattern. The bike did not alter public mores or fashion on its own, any more than it could do nowadays. At present women in strict Moslem societies, for example, do not cycle; in Saudi Arabia they are forbidden even to drive cars.

By the turn of the century High Society ladies were beginning to value the motorcar as a less effortful means of self-display, though motoring as a pastime for women on their own was not possible until the invention of the self-starter in 1911. And by the end of Victoria's reign the bike was totally accepted as personal transport for women of all classes, who still wished to pedal their way to freedom. So much did the bike become women's transport that it might fairly be asked whether in fact it was not women who liberated the bicycle, rather than the other way round. Even the search for the Safety must in part have been motivated by the manufacturers' realization that their market might be doubled overnight by the appropriate design. In the same way, sociables and tricycles, often made with the female rider in mind, had been the spearhead of engineering progress after the Ordinary had arrived at its mature development.

MODERN TIMES

By 1900 the bicycle had earned a place as an integral part of the transport system of the developed world. Its use had become for many a habit rather than a thrill. The cycle industry stabilized and devoted itself to producing a sensible range of machines. The high quality of the best models produced up until the thirties makes them collectors' items nowadays.

As we have seen, in the overall pattern of transport history the bicycle's most important role was, ironically, to prepare the way for the automobile. The petrol engine was evolving at exactly the same time as the bike achieved its modern form. Daimler began manufacturing cars in Germany in the late 1880s and by the turn of the century the clatter of motor exhausts had been heard in many parts of Europe and America.

Few people in Britain in 1900, 1910 or even 1920 could have foreseen with any certainty the vast success of the motorcar. To begin with cars must have seemed as rare and unobtainable as private aeroplanes nowadays, and it was only the very rich who took them up. Thus the car did not greatly alter the broad pattern of cycle usage in Britain until well after the First World War. Things were different in America, where Ford had begun producing autos for the mass market as early as 1903. In 1915 nearly three quarters of the cars made in the USA cost less than £200, and by 1919 the American motor industry was producing more then two million vehicles a year. But in Britain, in spite of the many car firms

Two quality machines from 1905: the Singer-de-luxe (top) and the Dursley–Pedersen.

begun by famous names between 1900 and 1914 (in 1913 there were 198 different models being produced), there was no cheap car on the market. The war saw both car and cycle firms turn to making armaments, so that car design and production stagnated.

It was the early twenties before the Austin Seven and the Morris Cowley made the breakthrough in Britain, providing cheap cars which created the expectation of family transport for a large proportion of society. For the next few years cars improved technically, while prices fell and incomes rose in real terms. A car costing £500 in 1922 could be bought new for £325 in 1926. The bike was doomed to lose prestige, at least with the middle classes, against this kind of competition. Numbers of cars and commercial vehicles, as well as motorized public transport, grew steadily, until by the outbreak of the Second World War there were well over three million mechanically propelled vehicles in Britain, two thirds of which were privately owned. Car-ownership was no longer a luxury for the few, though the great expansion was still to come after 1945.

'. . . the expectation of private transport for everyone'. Going to see Amy Johnson land at Croydon Aerodrome in 1932.

In the years between the wars it had been different. By the thirties the bike had taken up its role in an integrated transport system as the workhorse of postmen and policemen and the characteristic vehicle of manual workers and housewives. At the factory gates the bike racks were crammed full. Cycle racing, along with football, pigeon racing and coarse fishing, had come to be regarded as a predominantly working-class activity and had a large following. And the bike was still a much-loved instrument of leisure for many of its users. A week's holiday all-in at CTC hotels could be managed for £3 10s., and for £2 at the new youth hostels, based on German models and started in 1930. In 1931 not one family in ten owned a car. There were probably about ten million cycles to the one million cars in private hands. In the early thirties, at least, cycling clubs proliferated, but cycling enthusiasts had to contend with the more or less passive entertainments which were steadily gaining in popularity. The massive cinemas were packed nightly, in spite of the new wireless at home. The craze for swing filled the Palais de danse

1931: a ladies' cycling club in the winter sunshine of Epping Forest.

in the later years of the decade. Society was moving towards the mass leisure which characterizes the post-war years.

A 'weekend' in the open air was still an essential for many. In *The Uses of Literacy* Professor Richard Hoggart describes the activities of the club cyclists in a way that recalls the earlier observations of Flora Thompson, though without her freshness of detail. He mentions the thirties craze for hiking and 'getting out into the country in general':

> . . . the working classes went too, on to the dales and hills and moors, which luckily are not far from most of the large towns. If walking is not markedly typical of working-class people, then cycling is. A sign of arrival at real adolescence is the agreement from one's parents to the buying of a bike on the hire-purchase system, paid for out of weekly wages. Then one goes out on it at weekends, with a friend who bought a bike at the same time, or with one of those mixed clubs which sweep every Sunday through town and out past the quiet tram terminus. Many young people insist that they 'like their bed' on a Sunday morning, but a good number are out in this way. The membership figures of the two main cycling clubs give no real indication of the numbers who are out, but rather of those who have more seriously taken up cycling; yet even they have a quarter of a million members. For those who want club companionship, exercise, 'a good day out', there is the Cyclists' Touring Club (and many another local club whose members are not in the CTC); and there is the National Cyclists' Union for those who go in for racing, with specially selected tubing, wheels, and saddles, and an aluminium bottle in a cage on dropped handlebars. The NCU members may often seem scarcely to know whether the road passes through a town or a National Park: but sightseeing is not their purpose. The CTC members, talking as they ride or playing with a ball on the grass, seem to pay little attention to the countryside or the ancient monuments they visit. But they get what they come for – companionship, hard exercise, and fresh air.

Hoggart's armchair overview mixes personal reminiscence with wildly inaccurate estimates (the figures he gives for membership of the NCU and CTC more than double the actual numbers for all club-affiliated cyclists). He concludes that cycling offers 'valuable evidence that urban working-class people can still react positively to the challenge of their environment and the useful possibilities of cheap mass production'. Talk to any rider who remembers pre-war

Sunday rides, the best local spots for drumming up a cup of tea, the beauties of the hand-built machinery and the emptiness of the roads and you will get a much more vital impression of how it was than from Hoggart's faintly patronizing comments. (It is striking that a work purporting to be about 'working-class culture' should pay so little attention to sport.) But even as Hoggart was writing, in the late fifties, the activity he was discussing was dwindling and almost dying of neglect. Whereas the great age of cycling had still to come for the clubmen of Flora Thompson's account, Hoggart depicts behaviour which was fast becoming rare to the point of eccentricity.

The nadir of cycling arrived during those years when the population was continually being told that it had never had it so good. For

Cycle clubs boomed in the late forties, with most members in their teens and twenties. This young family rides a triplet and everyone takes a share of the pedalling.

the first time cars were generally within the financial grasp of the working class. The motor industry boomed. A new generation of drivers had been trained in the army, and full employment and hire purchase gave them the chance to buy a car. In the late forties and early fifties Reg Harris's world championships had made news in Britain and brought thousands to the cycle tracks. He featured frequently as a sporting hero in the pages of the *Boy's Own Paper* and the *Eagle*. By comparison, when Tommy Simpson became World Road Champion in 1965, the general public was largely uninterested. That same year there were nearly eleven million cars and motorcycles in Britain, roughly one for every five of the population – men, women and children. The number of cars had doubled in only a decade. There was much less room on the roads for the bicycle.

Yet it was the descendant of the Safety which had led this extraordinary transport revolution in the first place. It had fought many of the car's battles in matters of legislation, road usage and behaviour. The CTC had pioneered signposts and campaigned vigorously for better road surfaces. It persevered in putting up danger notices at the tops of hills until the Motor Car Act of 1903 provided for the placing of road signs by county councils. CTC enthusiasts formed a powerful lobby for cycling, showing the way for the motoring organizations which sprang up around the turn of the century. The League of American Wheelmen was founded in 1880 on the lines of the CTC, and they initiated conventions for hand signals in order to cope in encounters with horses. They too agitated for highway improvements, for even the city streets were in a sorry state. The bicycle manufacturers were happy to support them in this campaign. The very marketing and advertising techniques of the motor industry had been first developed by cycle manufacturers: the pomp of the 'Cycle Show' and the unveiling of 'this year's model'. And, of course, many car-makers began by building the two-wheeler. But these are trifles in the end, compared with the truly revolutionary stimulus provided by the bicycle. This was the urge towards a special freedom of personal travel, the urge to increase the perimeter of one's experience, quite literally, by undertaking journeys that only a few decades before would have been unimaginable.

THE BICYCLE IN WAR

The rise and fall of the bicycle's prestige as a machine of war provides us with a final model for its place in society at large during the last hundred years. The variety of uses to which the armed forces have – and have not – put the bike crystallizes its history, from the tremendous excitement it created during the boom years, to the equally marked lack of interest during the more technologically snobbish times that followed. Its role in war also has something important to tell us about what kind of machine the bicycle is.

Not surprisingly, almost as soon as the bicycle was invented speculation began about its possible use in war. In Britain as early as 1888 an army unit consisting wholly of cyclists had been formed, the 26th Middlesex (Cyclist) Volunteer Rifle Corps. Books of military drill for cyclists were written. America saw similar trends. The industrious Colonel Pope marketed his Columbia Light Roadster, quite falsely, as 'The Soldier's Standard Bicycle'. American cyclists ran messages for the National Guard during various strikes in the nineties. But when America went to war with Spain in Cuba the aspirations of the cyclist soldiers came to nothing, the terrain being absolutely unsuitable. Interest in cycling waned sharply in America

MORE ECONOMY.

A HINT TO "GOVER'MENT." A CHEAP REMOUNT FOR LIGHT DRAGOONS!

Punch, 1874

after 1899, largely because the automobile was so quick to establish itself there.

In Europe the excitement lasted longer. The cycle clubs, with their uniforms and bugles, converted readily into part-time military organizations. Companies of volunteer cyclists multiplied at a formidable rate. By the outbreak of the First World War there were no less than fourteen battalions of cyclists. The official army attitude to these cyclist-soldiers became a topic for controversy. In 1902 H. G. Wells, writing in the *Fortnightly Review*, angrily attacked the manuals for military cyclists, with their outdated notions of military decorum and sterile arguments about whether the bicycle brigades should be considered as infantry or cavalry. Army disciplinarians had suggested, for instance, that the gears of the soldiers' iron steeds should be standardized so that groups of cyclists could pedal in time. Absurd tactical manoeuvres were practised, such as the 'zariba', in which a hollow square was built of bicycles: the wheels might be spun so as to frighten the enemy's horses.

Though Wells mocked the War Office handbooks of drill, and spit and polish for cyclists, he was far from unsympathetic to the basic idea. The bicycle, he argued, 'may be destined to be the

"Bicycling News" Cartoon.

APRIL 30th, 1887

THE FUTURE OF THE WAR CYCLE.

dominant arm in the European warfare of the future'. His short story, 'The Land Ironclads', describes the future war he envisaged. Townsmen with minimal training, but equipped with 'iron-mongery' in the form of bicycles and primitive pedal-driven tanks, outflank and easily capture an apparently superior force of crack cavalry and infantry, whose complacency gives way to surprise and indignation at being defeated by an amateur force of social inferiors. The story is, indeed, a science fiction version of what had already happened with the Boer horseback commandos in the war in South Africa. Wells was no lonely eccentric in his views of future modes of warfare. In Italy almost the entire circle of Italian Futurist architects, poets and painters joined the Battaglione Ciclisti Lombardi at the start of the First World War.

The bicycle proved itself as a useful accessory in both world wars, but not at all in the way Wells had imagined in his fiction. In the Second World War, folding bikes were used as special equipment for Commandos infiltrating behind enemy lines. But, in general, far from being the glamorous spearhead of attack, it became just another piece of equipment to be utilized at will. Bicycles were

Belgian Congolese cycle troops on parade in the desert during the Second World War.

Pedal power, wind power: a folding bike is loaded on to a glider for the Rhine crossing.

An identity check as refugees flee from Dunkirk during a short truce in the 1944 Allied invasion.

issued to speed a landing or advance and then soon abandoned; they were a good means of moving men along roads when not under attack, or when petrol was scarce. And, of course, a stolen bike was the tired infantryman's best friend. Families of refugees moved across Europe on dilapidated bicycles, pulling their belongings behind them in hugely overloaded trailers. On the Home Front, cycling thrived.

The bicycle could still have a viable role where particular terrain or circumstances favoured its use, but such circumstances were not always obvious. The capture of Singapore in January 1942 serves as something of a parable about the dangers of neglecting or forgetting the capabilities of this formidably adaptable machine. Before the war bicycles had been one of Japan's chief exports. When the Japanese decided to attack Singapore, their army took thousands of bicycles with them onto the Malayan peninsula. Each company of soldiers included an expert bicycle repairer. As the troops advanced

Two young German soldiers left behind to ambush tanks. Laden with anti-tank rockets and supplies, their bicycles have become unobtrusively powerful weapons. But the riders look very vulnerable.

they seized more cycles from the local villagers, knowing that they would be Japanese models and that all spare parts would be interchangeable with their own machines. The British, with highly accurate maps, regarded the Malayan jungle as impenetrable; the Japanese troops, who only had school atlases to guide them, simply cycled down the long straight roads to Singapore. Whereas a truck, tank or armoured car would have been stopped by the bombed roads and blown-up bridges, the Japanese soldiers hoisted their bicycles onto their backs and climbed over the obstacle or forded the river. In the midday heat bicycle tyres burst and, not having time for repairs, the troops cycled along on the wheel rims. The noise of many hundreds of steel rims grinding and drumming in the darkness sounded to the nervous and bewildered defenders just like the rumble of advancing tanks, and they fell back continually before an imaginary superior force coming towards them with incomprehensible speed. When the Japanese army arrived at Singapore they found the British guns fixed pointing out to sea. The rear defences and the jungle had seemed impregnable.

Thus one part of the vision of 'The Land Ironclads' was proved to be accurate. And the episode of Singapore has been shown not to have been a unique and mildly farcical tactical aberration. The presumption of Western culture that technical superiority is invincible was disproved again and again in Indo-China, first for the French and then for the Americans. Peasants with bicycles have endured the most expensive and powerful war machinery known to man.

The family returns to a war-shattered town in Vietnam.

THE PRESENT

For the last three quarters of a century the bicycle has been available as a simple, cheap and efficient means of transport. It is broadly true that it has been neglected and forgotten in proportion to the economic sophistication of different areas in the world. The more highly developed the country, the more it is likely to rely on expensive power sources rather than human muscle.

Decoration and function go together in this Indian rickshaw with its flowers, flags, bells, inlaid panels and elaborate upholstery.

In the so-called Third World, the prestige of the bicycle differs radically from its standing in the West. China provides the most obvious case. There private car-ownership is forbidden by law, though important party officials use cars and jet planes. In China even women cycle, which is uncommon, for example, in most of Africa. The Chinese export their New Phoenix and Flying Pigeon machines all over the world. In India the city streets are monotonously noisy with the continuous din of bicycle bells, part of the constant war of nerves between the cyclist and the pedestrian. In Africa too, in spite of the climate, the bike is a highly prized form of transport, especially in the many areas where the tsetse fly kills all beasts of burden. Every West African ice-cream seller or market porter with his wheelbarrow or *pousse-pousse* no doubt dreams of owning a moped or a taxi, or even a Mercedes like the politicians;

Pedal power in Peking: three bicycle trucks and (opposite) a waste-paper collector make their rounds.

but however the elites of the Third World are entranced by Western values and possessions, the two-wheeler still allows its owner a freedom and an earning power denied to those who have to walk.

There is a vital difference between the use made of the bicycle in the West and in the underdeveloped countries; it is often overlooked by those radical dreamers who make special claims for the bicycle as an answer to the transport problems of the overdeveloped nations. The distinction takes us back to the essential factors which provided the foundation for the 1890s cycling craze, those factors with which this chapter began.

In the Western World the bicycle has always been, and still is, predominantly an instrument of leisure. Its convenience and utilitarian value have been at least matched by its potential for pleasure. In the Third World the emphasis is different. To begin

with, traditional societies do not share our concept of leisure. Leisure itself, like the bicycle, is a comparatively late development of urban industrialization. For someone whose working life is not split into stretches of work broken by occasional rewards of permissible idleness, the bicycle is primarily a tool, a beast of burden, rather than a luxury. In Europe and America the history of the bicycle is linked with the development of the idea and the organization of leisure, the steady rationalization and standardization of working hours, the legal enforcement of days off, yearly paid holidays, and so on. In Britain Sir John Lubbock's Bank Holidays Bill of 1871 began the drawn-out legislation to make statutory the gradual relaxation in working hours which had been initiated by the spread of the sixty-hour week during the 1860s. Organized sport, hobbies and tourism all required the leisure which resulted from successful industrialization. It is no mere coincidence that the spread of cycling occurred at exactly the same time as the foundation of the associations which govern tennis, cricket, football, hockey and all the other major sporting activities.

At present the bike is in fashion again, and the differences in attitude to it in the overdeveloped and underdeveloped countries of the world are useful pointers to the nature of this revival. Is the interest in cycling, started in the seventies in the United States and spreading to Britain in the following years, no more than a passing craze, a shortlived bonanza for the bike industry? Or will the bicycle become a properly valued mode of personal transport, and even recover some of its past glory?

5 A Visit to the Industry

The bicycle may be news again, but it is no stranger to periods of boom and slump. The mass production of the Safety created gold-rush conditions, and by 1897 an incredible 830 or more manufacturers were at work in the main industrial cities of Britain. By the same year the newer but faster-moving American industry had grown to 700 cycle-making companies which had begun to compete on the export market as well. It is small wonder, then, that the boom soon produced so many advances in production and such fierce rivalry that within ten years companies were having to cut their profits more and more drastically in order to stay in business. Many did not survive and a list of deaths, mergers and shotgun marriages among bicycle-makers of the day would keep a genealogist happy for hours.

In the years before and after the Second World War the two-wheeler may have had an 'austerity' image, but cycling nevertheless thrived as transport and pastime. Curiously enough it was in the late fifties that the real recession came, and this slump hurt the industry badly. Production at Raleigh fell by as much as fifty per cent during this period of contraction. Everyone in the business refers to those times with something of a shudder. The causes are not easy to identify, but among them were increases in prosperity, hire purchase and the growing ownership of small cars. Whatever the reasons, the eight years following 1957 were bleak ones for bicycles and many British suppliers and companies folded up or went over exclusively to the motor industry. In other cases it was only foreign sales that kept businesses going. For example, without their French market, Reynolds would probably have abandoned cycle tubes at this time.

The hard times passed, however, and from the mid-sixties trade began to increase and the production of small-wheelers and a growing leisure market in America stimulated a new interest. In 1973 over thirteen million cycles were sold in the United States – more bikes than cars that year – and dealers could not buy enough quality lightweights, especially those with prestigious English and European names. It was a boom again, although a more modest one than the explosion of sixty years before. American sales have since slowed down and it seems likely that everyone who wanted a bike has now got one. Although enthusiasts will keep changing mounts or collecting them, and although there will always be up-and-coming young riders, the new popular market has probably stabilized. After all, sales saturation of this kind is not unique. Thirty years before the first bicycle rush there had been a similar expansion in the sales of sewing machines, and the same process has been seen in action today with the spread of cheap pocket calculators.

The problems of marketing are intriguing and, for ourselves at least, the business of designing and building bicycles is even more interesting. So it was that we wanted to take a look at the state of the art in Britain today. We could not approach every company, of course, but we wrote some letters, visited dealers and talked to anyone who would listen to us. People were kind and we were lucky to be able to satisfy the curiosity that most riders must feel from time to time as to exactly how their machine came to be made and what the place that built it looks like.

When we went to Birmingham and Nottingham, British bicycle sales had been rising steadily but unspectacularly over the last three years and there was a general feeling in the trade that at last cycling was beginning to catch people's imagination as a practical and healthy means of transport and recreation. The people we met were confident that the increase in business was here to stay, even though the astounding expansion of the American market had clearly slowed down. Yet the ghost of the late fifties was still around, squeaking somewhere in the corners of the boardrooms, and men were cautious with their predictions.

The industrial Midlands have a long history. In the late nineteenth century, Birmingham, Nottingham and Coventry were

LONDON DEPÔT—5, HOLBORN VIADUCT. 11

THE "ROVER."

(REGISTERED.)

IMPROVED FOR 1891.

The First Machine to accomplish over 21 Miles in the Hour, Bordeaux, August, 1888.

SPECIFICATION.—30in. wheels, geared to 54in. (unless otherwise ordered); direct spokes; adjustable ball bearings to both wheels, crank-axle and pedals; tires, ¾in. to front and ⅝in. to back wheel; adjustable handles and seat-rod; plunger spoon brake; wheel and chain-guards; best saddle and spring; plated lamp bracket, foot-rests, etc.; valise, spanner, and oilcan; finished best black enamel, and lined in two colours; all bright parts plated. Also made with 33in. front wheel.

PRICES.

Ordinary Tires (¾in. and ⅝in.)	£20	0	0
Black Tires (⅞in. and 1in.) Special	21	5	0
Cushion Tires (1¼in. to both wheels)	22	15	0
Pneumatic Tires	25	0	0

Extras.—Gold lining, 15/- Crates, 2/6 each, not returnable.

TELEGRAPHIC ADDRESS—"LOCOMOTION, LONDON."

where workers began on the shop floor and ended as aldermen at the heads of the leading families of British industry. It was a golden age for confidence, speculative nerve and native engineering genius. From the viewpoint of the 1970s the unselfconscious success of these dynasties seems as distant, as romantic and in some cases as piratical as the career of Francis Drake. Take, for example, sixteen-year-old James Starley, who ran away from his peasant home near Brighton and twenty-four years later, in 1870, had founded his own cycle-making business in Coventry. This is no fairy story, nor was his case unique. An early partner of his was one William Hillman, an ex-foreman at the Coventry Sewing Machinists' Co. where Starley had worked on the shop floor with another friend called George Singer. Hillman and Singer, of course, were to go on to make motorcars bearing their names.

These inventive times are delightfully characterized by the story of how James Starley perfected the differential hub, as told by Geoffrey Williamson in *Wheels Within Wheels*. James was riding with his burly son William on a Starley machine known as the

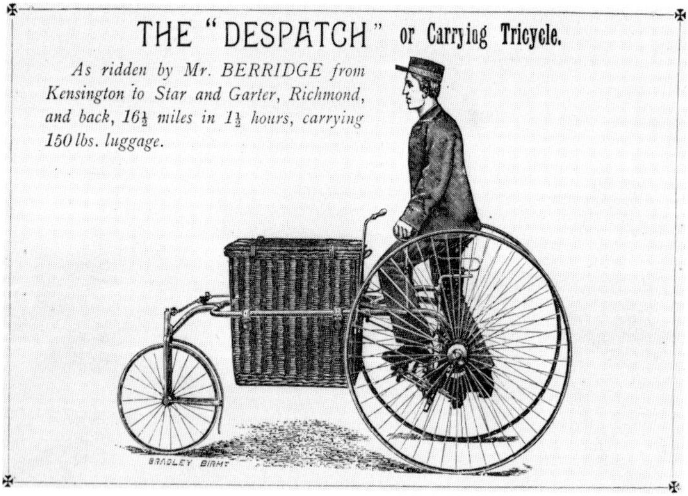

The 'balance gear' was also fitted to Starley's Despatch tricycle, 'constructed to carry any kind of package or parcel (any reasonable weight), such as Grocery, Drapery, Milk, Fruit, Vegetables, Bottled Beer, Wines or Spirits, small barrels of ditto . . .' The price was £25.

Honeymoon Sociable. The cycle had two saddles and a common crank, so that it looked, in fact, like two Ordinaries joined together side by side with the riders sitting between the big wheels. It was young William's stronger pedalling on a hill which swerved the machine and tipped his father into a ditch. James sat by the roadside and thought hard for a while, identified the problem and hit upon the modern differential gear. Later, at home, this remarkable self-taught engineer realized his inspiration, as he did with most of his inventions, by carving a model of it. Jeering onlookers were never far from the scene of a mishap in those days, but they did not recognize what they had really witnessed when James Starley fell off, his vanity hurt, his hand stung by nettles.

The Honeymoon Sociable was short-lived, but when James's nephew produced the Rover Safety in 1885 it was to set the design for the future at a time when there were some two hundred companies still making the high-wheeled Ordinary. Development moved fast in those days. Only three years later, for example, Frank Bowden founded the Raleigh Cycle Company in Nottingham. Bowden was a wealthy lawyer in his late thirties who retired from business in Hong Kong because of poor health. Back in England, a doctor advised him to take up cycling. He was so impressed by his recovery and by his two-wheeler that he bought the company which made it, calling it after the firm's address in Raleigh Street. This must be one of the cycling health-cure stories of all time. In workmanship and materials some of the machines of the early 1900s have scarcely been surpassed. Humber, Sunbeam, Singer, Raleigh and Dursley-Pedersen were all associated with the best in quality.

Frank Bowden was impressed by the Changeable Gear on his Woodhead, Angois and Ellis cycle, and it continued to feature on the newly named Raleigh.

These were the pioneering days when the business was filled with a bewildering variety of independent makers and endless patented accessories. Today, however, the industry is more centralized, not to say polarized. There are the relatively 'hidden' firms of Halmanco and Comrade, which build bicycles for Halfords and for the warehouse trade. There are a few small independent manufacturers. Then there is the industrial giant of Tube Investments Ltd, which contains dozens of companies, including British Aluminium, Reynolds Tubes and firms which make cooking foil, gas water-heaters and spin dryers. TI acquired its cycle-makers over the years, absorbing the British Cycle Corporation, Hercules, Phillips,

Norman, Sun, Armstrong, James and Aberdale. The consortium's greatest competitor in this field was Raleigh Industries, which had, in its turn, incorporated Rudge–Whitworth, BSA, New Hudson, Sunbeam, Triumph, Three Spires, Humber and Carlton. Finally, in 1960, when the cycle industry was in such poor shape, the two competitors merged for mutual support and became TI–Raleigh, with Raleigh handling the entire cycle side. The pattern is a familiar one in modern business, but it seems a pity that in so many cases all that really remains of those veteran independents is their names.

Working in the style of Beardsley and Art Nouveau, Will H. Bradley was the best-known poster artist in the United States at the turn of the century. In 1887 it was A. H. Overman's Victor machines which ushered in the American boom in Safety bicycles.

To help explore the modern cycle-makers we divided the field into four elements. Firstly we wanted to visit Reynolds, who make lightweight tubes for the whole industry and export them to Europe and America as well. For many manufacturers, both big and small, quality production begins here. Then we wanted to see T I–Raleigh, the largest cycle producers and exporters in the world. (In 1972 T I–Raleigh's advertising budget was nine times that of all the other British cycle-makers put together.) In contrast our third company was Dawes Cycles Ltd, a more modest business, independent and apparently flourishing. We wanted to find out how they managed in a highly competitive market. Other similar companies are Falcon Cycles of Humberside and the recently revived Viking Cycles in Northern Ireland. Finally, there are in Britain a large number of specialist lightweight frame-builders, most of whose names are known only to the *aficionados*. Yet it seemed to us that these tiny businesses still do much to keep alive the reputation of the British hand-built lightweight. Out of the dozens we might have visited we decided on a newcomer to the scene, Dave Moulton of Worcester.

REYNOLDS

Every quality British bicycle could be said to begin at the Reynolds works in Tyseley, but the first thing we saw on our visit was a mummified cat and some ancient roof nails. They had been found during the renovation of Hay Hall, a fourteenth-century manor house surviving rather incongruously in the middle of the works. Having preserved Hay Hall, the company had also decided to keep the cat and nails, for luck perhaps, in a glass case. We were to learn later that the Reynolds family began by making nails in 1841, and so their interest in those antique iron spikes is understandable. (Indeed, John Reynolds and Sons Ltd still make nails in West Bromwich.)

Reynolds' association with bicycles began when they decided to expand their business and young Alfred Reynolds, one of the founder's grandsons, hit on a plan to produce seamless steel tubing for the booming cycle trade. Builders want lightness and thin-walled tubes provide it, but the problem is that they may not be sufficiently strong at the corners of the frame where they join together and take most stress. Young Alfred and J. T. Hewitt solved this in 1897 by patenting a process for making butted tubes which had internally thickened walls at each end. Thus they were strong where they needed to be and light as well. The Patent Butted Tube Company was registered the following year and began to supply the industry with its high-quality materials.

Nowadays cycle tubing is only a tiny part of the Reynolds organization but they still treat it seriously and take pride in a programme of continuous development. Successive years saw AA quality, then Reynolds HM, and finally, in 1935, the first version of the now famous Reynolds 531. The figures 'five, three, one' represent the broad proportions of some of the elements that go into this manganese–molybdenum alloy steel. The formula has been further improved over the years and it is one of the strongest materials for light construction in the world. More recent developments are a Special Lightweight set of 531 tubes in thinner sections and Reynolds 753, a new alloy steel so strong that it allows exceptionally light tubing to be used.

At the beginning it was Reynolds' expertise with lightweight cycle tubes that led to contracts for aircraft materials in the First World War. This turned out to be a most important new direction because the growth of civil aviation and the expansion of the R A F meant a whole new market. Today, in addition to tubing, the Company produces integral-ended hollow extrusions for rocket motors, oxygen cylinders, railway buffers and hydraulic cylinders. They also make flash-welded rings in titanium and extremely costly alloys for jet engines throughout the world, including the Rolls-Royce engines for Concorde. In turn, it was largely this aero-engineering experience which led to the refinements of steel alloys for 531 and 753 cycle tubes. 531 tubing is also used in other products such as aircraft seats, wheelchairs and stretchers, and was at one time used for the engine mountings of the E-type Jaguar.

In contrast to Reynolds, the huge Birmetals plant which we visited at Quinton, in Birmingham, came to cycle parts only recently and from the opposite direction. As Europe's largest supplier of magnesium alloys, after thirty-five years in the aircraft industry, they brought their considerable aerospace experience to the making of light alloy handlebars, wheelrims and seat pins. Accordingly, the Birmalux division now actively sponsors cycle sport, and in cooperation with British Leyland provides a service vehicle for the Milk Race with a crew and a roof-rack full of spare wheels. Since the

1970s Birmalux has supplied bike-makers like Falcon, Trusty, Dawes, Holdsworth and Comrade Cycles but, as with Reynolds, this accounts for only a small percentage of a business which sends irrigation pipes to Cuba and Romania and Land Rover bodies all over the world.

At Quinton and Tyseley we saw once again how often nowadays the absolute priorities of strength and lightness link aerospace technology with the materials to be found in a racing bicycle. Thus it is full circle for Orville and Wilbur Wright, those cycle-shop mechanics and dreamers. Could they ever have predicted that, some seventy years after Kitty Hawk, Signor Tullio Campagnolo would advertise his beautiful cycle hubs by modestly admitting that his company also makes bits for satellites?

In Reynolds' cold drawing section our immediate impression was of a soapy, hot-oil smell in a huge, high, glass-roofed shop. Everywhere there was the distinctive clang of steel tubes being stacked or moved by overhead travelling gantries. The process begins when 'hollows' or 'blooms' are brought to the draw benches. These already look like tubes, having been hollowed out at the foundry while the metal was still hot. These blooms are first of all softened by heating (annealed), pickled in acid to remove scale and then lubricated with a special mixture of oil and soft soap. They are now ready to be pulled through special dies into longer and thinner-section tubes, and this is what is called cold drawing. We saw how powerful electric motors at the end of each draw bench pull the tubes slowly and smoothly through the dies. Six or seven passes may be made before the correct gauge is attained and the tubes have to be annealed and treated again between each pass because the drawing process increases the tensile strength of the steel each time. Some of the benches therefore save effort by handling as many as six tubes at once, each bench under the control of a single operator. We moved in for a closer look.

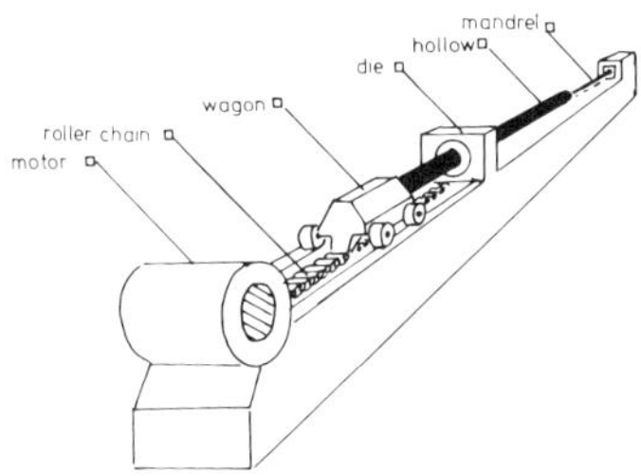

Fig. 17

Each draw bench has a long track and an endless roller chain to which a wagon is attached. The wagon has serrated jaws (called 'dogs') which are there to grip the 'tag end', the pinched-in extremity of each bloom (Figure 17). The process begins when the

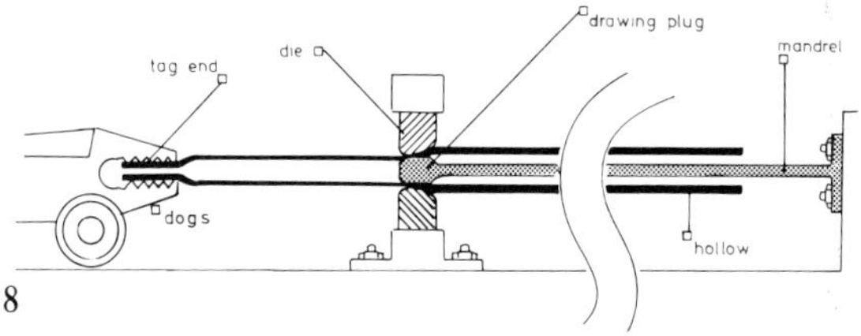

Fig. 18

hollow is threaded over a mandrel fixed to the far end of the draw bench – like a giant drinking straw being slipped onto a horizontal knitting needle. The mandrel has a plug (which is in fact the die for the internal diameter of the tube) and this comes to rest against the narrowed tag end. The tag end is then pushed through the die at the middle of the draw bench and gripped by the dogs of the wagon on the far side. The wagon connects with the chain under the bench and moves off, the plug takes its place in the middle of the die and the slippery tube is relentlessly drawn down to section between them (Figure 18). Some of these draw benches are nearly a hundred feet long and all sorts of tubes and sections can be made on them with gauges accurate to within three thousandths of an inch. Seamless tube is particularly suitable for highly stressed mechanical purposes or for work under pressure and heat, as in boilers, chemical plant and hydraulic systems.

We left the clangorous rows of benches and cranes and walked to a quieter end of the building where the special cycle tubes are cut to length and finished. On the way, we passed walls of crates packed with completed tube sets labelled for America, Holland, Italy, Belgium and France. (65 per cent of Reynolds' cycle-tube business goes to the builders of France.) Then, in a space the size of an average school gymnasium, we found the men and the machines which make a collection of tubes into matched and numbered frame components. They enjoyed puzzling us with their expertise.

If the mandrels on the draw benches had made us think hard, their behaviour in the double-butting process seemed little less than magic. The principle is simple. A shaped solid steel mandrel of specific length and section is slipped inside a slightly longer 531 tube (already cut to length and destined to be a top tube or a down tube on some completed bike). The steel is then passed into special rollers

Tubes at the die face. The rig draws three at a time and then rotates.

and through a die which draws the tube down around the mandrel, now trapped like a solid sausage-filling inside it. This gives the tube the thin walls and thicker ends characteristic of the double-butted process. (The 19/22 gauge changes from 1 mm thick to 0.7 at the walls, the 21/24 gauge from 0.8 mm to 0.5 mm.) So how do you get that solid steel mandrel out? 'Simple,' said Cyril Barlow. 'You pass it through these inclined high-speed rollers which spin the tube and

Butted cycle tubes being tapered.

make it spring the mandrel out.' We saw it happen and did not believe it.

At other machines the thinner tubes for chain and seat stays are tapered and waisted to a variety of designs and lengths according to the preferences of the builder. As many as fourteen or fifteen operations go into the 'taper-gauge' fork blades, since they have to possess several special qualities. The challenge is to produce blades which are light, strong and resilient. The problem is that the tapering process would tend to leave the tube section thicker at the narrowed end of the fork (where resilience is wanted) and comparatively thin at the top end, near the crown, where most strength is required. Reynolds solved the problem by butting the crown end of the fork blade and by making the tube walls of gradually thinner gauge towards the end that will eventually hold the axle. Then the tube is tapered at this end and the thinned walls more than make up for the 'thickening' effect of the taper. The blade is next formed into an oval section at the top and curved to any one of five possible rakes from one to two and a half inches (25–65 mm). (Track forks have a round section only.) The result is a lightweight taper-gauge blade, strong enough to take road shocks without deforming.

The heaviest tube in the whole bicycle is the steering column, only seven inches or so long and hidden inside the head tube. This part must be very stout and rigid, so it too is butted at the end to be brazed to the crown, so as to make the completed fork assembly as strong as possible. The famous rider Eugène Christophe knew about the importance of this, as his forks broke at crucial moments no less than three times in his racing career.

The first damage occurred in 1913 when he was leading the Tour de France on a descent of the 7,000-foot Tourmalet. Undaunted and only slightly injured, Christophe carried his machine with its broken fork column for five miles down the mountain until he came to a village blacksmith. For two hours the rider laboured at the forge until he had made the repair on his own. Those were the rules in the heroic early days – and even then he was penalized for letting a boy pump the bellows for him. The repair cost him four hours overall. After the war, when the Tour began again in 1919, Christophe was the first man to wear the newly introduced yellow jersey. Half-way through the race he was in the lead, until once again his forks broke and once again he repaired them himself. This time he finished third overall. Christophe rode his last Tour in 1922 at the age of thirty-seven, only to have his bike break once more on the rough Alpine stages. He still finished the race. 'Le Vieux Galois' never won a Tour out of the eleven he rode.

We began to see why Reynolds take such pride in the strength of

Forming fork blades to a pre-set rake.

their product and why the coveted green and gold transfers cannot be bought over the counter. People have been known to snip pictures of them out of advertisements and paste these little paper squares to their old clunkers. There was even a 'forged' transfer created for a joke in America, which on close inspection read: 'Guaranteed not built with Reynolds 531'.

There are only a few lightweight butted cycle-tube manufacturers in the world and they are all highly regarded by framebuilders. In France, on the banks of the Loire at Unieux, Ateliers de la Rive make Durifort for strong sporting and touring use, as well as Vitus 172 and the even lighter Super Vitus 971 for road racing and track cycles. Columbus butted tubing is made in a variety of gauges by A. L. Colombo of Milan, and it was their superb Record set that was used when Eddy Merckx broke the world record for the hour on the track in Mexico in 1972. (This machine was built by Ernesto Colnago of Italy; it weighed only twelve and a half pounds complete and when it appeared at a cycle show recently it was insured for £2,000.) In Japan, Ishiwata, Tange and Day all make chrome molybdenum double-butted tubes, some of them also very light for special-purpose record attempts. Reynolds is high among these prestigious names and the recent developments in 531SL and especially in 753 have done much to keep them internationally sought after. A 753 frameset weighs under 4.4 lbs (2 kg); the walls of its tubes are an incredible 0.3 mm thick, and yet they are half as strong again as standard 531, with a tensile resistance of 75 tons. Undoubtedly the green and purple 753 transfer is a very superior badge indeed: a long way from that nail factory in 1841.

The largest bicycle plant in the world, Raleigh Industries at Nottingham produces over two million bicycles a year.

RALEIGH

To visit the Raleigh Cycle Company we had to turn north-east along the Trent valley to the city of Nottingham. The Raleigh factory is well within the city bounds and the main building on Lenton Boulevard presents a modest neo-classical façade, rather like a public library of the 1930s. In the foyer, complete with uniformed attendant, there are displayed racing trophies, maps of the Company's foreign holdings and a tall revolving chrome sculpture welded together from miscellaneous cycle parts. Then one realizes that this is only the office block and behind it there lies a vast sixty-four-acre wedge of no less than four factories with miles of roads and conveyor lines. This is by no means the only Raleigh site in the country, but 8,500 of the Company's 10,000 British employees work here in Nottingham. Around the globe in India, South Africa, Nigeria, Holland, Ireland, Canada and the United States, Raleigh factories and subsidiaries produce over four million bicycles a year. It is a big business, the largest of its kind in the world.

Raleigh Industries Ltd is a holding company which includes Sturmey Archer Ltd, who design and make cycle and car components and who in turn include Brooks Saddles of Birmingham; there is a service division for accounts and purchasing, and of course the Cycle Company itself, which includes Cox of Watford, best known for their award-winning designs for car safety seats. These companies, combined with the resources of Tube Investments Ltd, mean that Raleigh can make almost every single part of their standard bicycles, from the machine tools themselves to gears and high tensile steel tubing, to cast alloy wheels and plastic mouldings for the toy division. The top models of Raleigh and Carlton sports cycles do use some alloy parts from other companies such as Weinmann, GB, Vagner and Simplex. These machines and the team Carlton–Weinmann racers are made at Worksop in the Carlton factory, while the racing cycles for the TI–Raleigh team in Europe and for Mid–Let–BSA in Britain are made in the Specialist Development Section at Ilkeston and furnished with Campagnolo equipment.

From cycles to components to toys, each division has concept and production design departments constantly at work. Raleigh export seventy per cent of their production and they have subsidiaries and retail outlets in more than fifteen countries. Colour schemes, decals, equipment and catalogues are all altered subtly to meet the needs of each market and be competitive in it. Even the names on the cycles are different, and we were surprised to see some of the old firms still appearing in shiny new metal badges: Rudge, New Hudson and Phillips. With over 140 markets abroad and some 6,000 different specifications to cope with, it is no wonder that the whole system of order and supply has been computerized.

We were struck by the size and variety of Raleigh's export trade. We were shown sports models for the United States complete with a high standard of alloy accessories; and a black enamelled 'Low Gravity Carrier' – the humble delivery bicycle with a small front wheel under a huge wicker basket. (The basket is an optional extra.) By far the Company's largest single foreign market is Nigeria, where there is a growing interest in sporting and recreational cycling. Raleigh professional riders have been there to hold seminars on road racing, even though the climate and the roads are difficult for competition – not to mention problems like snakes crossing the road in front of the bunched riders. Racing is still in its infancy in Africa, but reports from Nigeria and South

Pedestrians and cyclists mingle in a street scene from Lagos. There are no women riders.

Africa (where there is a black team) and the success of events like the Tour of Cameroun and the Tour of Zambia show that black riders may yet come to equal the world-class performances of black athletes.

The cornerstone of African cycling, however, is still the heavy-duty roadster, a tough workhorse of a bike that Raleigh make second to none. These machines captured our fancy. They are country cousins to the British Raleigh Superbe and their appearance has not changed for over sixty years, with very 'easy' frame angles, roller-lever brakes, chaincases, tool pouches and broad leather saddles perched on triple-wrought coil springs. We admired the roadster with two-inch balloon tyres, the Double Top Tube roadster with 28-inch wheels – very popular in the Middle East – and, most of all, the 'boss bike', an all-chromium-plated Superbe Roadster, a glittering status symbol. Gold carriage lines are still hand painted onto many of these export models, and indeed the whole trade displays a conservative taste that has almost disappeared on the home market.

A Low Gravity carrier and a Superbe Roadster. The sprung saddle, relaxed frame angles and 28-inch wheels on the roadster all contribute to strength, stability and comfort, but the all-chrome finish is pure style.

There is a long waiting list for a new bicycle in China.

These machines are shipped abroad CKD ('Completely Knocked Down') and twenty-five of them can be packed in two cases, each less than three feet square, ready for assembly in places as far afield as Nigeria, India, Mauritius, Malaysia or Ecuador. You can buy the chrome Superbe in France but not in the United Kingdom. That was a pity, we thought. The Chinese think enough of these roadsters to make and export at least two machines on very similar lines: the Swan Palace and the New Phoenix. The resemblance goes as far as the style of the badge on the head tube, except that the Raleigh crest is a heron, a bird with an altogether humbler role in mythology.

Building cycles at the Nottingham works is an impressive business because so many parts are made from scratch. Everywhere we went there were overhead conveyor lines taking a multitude of parts between buildings, from process to process, in and out of store, and finally routing them all together for assembly. Ramps and roads connect many different shops, and workmen and supervisors scoot between them on a variety of 'departmental' bicycles. Yet however much the processes are automated, making bikes is still a very labour-intensive operation. From the very first, men are needed to handle the jigs and to set up work on them, and this goes on from stage to stage until at the end the entire machine is hand-assembled on the lines. This need for intensive manpower faces all cycle-makers; it is an even greater imperative for the small firms, and it is these labour costs as much as anything else which dictate the price of the finished article.

Raleigh were committed to manufacturing large numbers of bicycles from the time they began production in 1889. Within ten years they were employing 850 people in the biggest cycle factory in the country, and with the help of their revolutionary immersion brazing method they were producing 12,000 cycles a year. Even by this time Raleigh had made a reputation for high quality and lightness. The famous tubular fork crown was introduced in 1892, and pressed steel lugs were developed soon after with the help of American machine-tool experts. Raleigh now advertised 'the All Steel Bicycle', which had advantages over its rivals still using heavy and brittle cast lugs. Frank Bowden also made it a policy to advertise racing successes, with his bikes ridden by the national amateur champions of France, Austria, Italy and Wales.

Perhaps Raleigh's most famous rider at this time was the American Arthur Augustus Zimmerman, amateur champion of Britain and the World and one of the first stars in professional cycle sport to use his name to endorse cycle products like shoes and clothing. 'Zimmy' enjoyed social life to the full and could still win sprints the next day pedalling a low gear at very high rates – a technique he learned on the racing Ordinary. Zimmerman was not alone, for American track-racing men were world contenders at the turn of the century. The glamour of this scene and Zimmerman's talent undoubtedly helped Raleigh's prestige; it seems that sporting successes have always been an important factor in marketing bicycles.

Indeed, the modern T I–Raleigh team's involvement abroad is a direct result of research in the late sixties which showed that brand awareness of the Company's products was declining in several European countries. (Holland, where Raleigh is represented by the 'Gazelle' name, is the exception.) The impact of sporting successes in

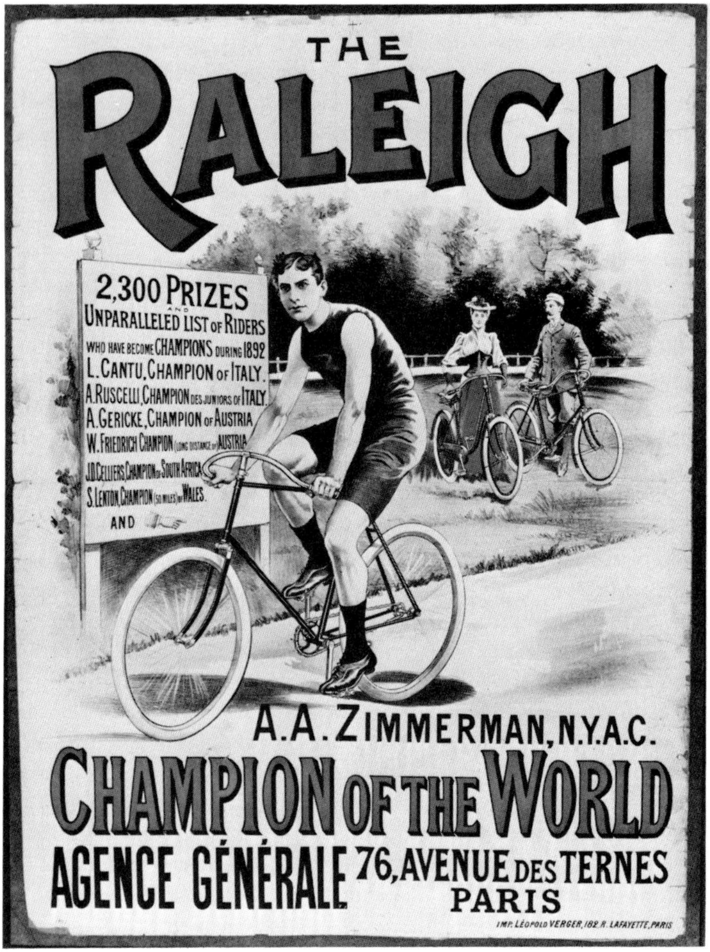

the fifties – the 'Reg Harris Era' – is emphasized in G. H. Bowden's book, *The Story of the Raleigh Cycle*, but recent times have been just as striking. By 1977, for example, the company had invested three quarters of a million pounds in four years of racing abroad. Their efforts were rewarded in the '77 Tour de France when the T I–Raleigh riders featured well as individuals and also carried

" RIVALS."

Sporting competition was commercially important from the very first. In 1884 Messrs Hillman, Herbert and Cooper organized a 100-mile road race for riders of their Kangaroo, and George Smith won with a record time for the distance. The next year he broke his own record – this time, however, in a race organized for Rover riders. Within a month the Kangaroos retaliated and the record fell once more to 'Teddy' Hale with a time of just over 6½ hours.

George Moore's drawing symbolizes this competition, made all the more intense, perhaps, by the fact that rival designers Hillman and Starley were once colleagues at the Coventry Sewing Machine Co.

off the team prize. In that same year the U K company was geared to export £12 million worth of cycles to Europe alone – a return of sixteen times the total racing budget.

Bike-building begins with the frame, and we saw hundreds of them being made from the forty miles of tubing which the frame shop uses each day. Children's cycles, small-wheel shoppers, roadsters and export 'jungle bikes' are all made up in their different ways. We saw tubing for Raleigh 20 and Chopper frames being carefully spot-brazed by hand, and we saw entire roadster frames being put together in only one and a half minutes on ring-brazing rigs. In this process the butt end of each tube has a ring of brass pinched onto it before being fitted into the appropriate lugs. The frame is then set up in a vertical brazing rig which revolves each lug in turn past fixed and roaring gas flames which heat the lugs and melt the brass inside to braze the tubes together. A completed frame is removed for quality checks every twenty minutes or so.

Immersion, or dip-brazing, is also used for some bikes and is almost as quick. Here the tubes and lugs are pinned together as usual and heated up in a gas flame before being dipped into a shielded trough of molten brass. The liquid metal penetrates all the joints at once and after less than half a minute the workman lifts it out, shakes off the excess brass and sets the piece aside to cool. At this stage frames were popped in and out of jigs to set and check their alignment, and finer adjustments were done later when the frames were cold. They were joining up head tubes when we stopped to watch, handling the glowing metal with speed and sureness in the glare and heat from the vats.

Methods like this, and the controlled gas flames on the ring rig, help to ensure that brazing temperatures are constant and evenly raised and lowered without any sharp changes. Excess flux and brass are removed by pickling in acid solutions, by washing and by using tools to grind the brazing smooth. Then the bottom bracket, head tube and seat tube are reamed out and threaded where appropriate before the frame is passed on to be shot-blasted before 'bonderizing' to provide a surface ready for paint.

Small batches are spray-painted by hand, but most bicycles and components are carried on moving hooks through automatic electrostatic spraying rooms. In this system the paint is poured onto a fast-turning horizontal disc which rises and falls as the frames slowly take a turn around it. The disc is given a huge positive electrical charge and the frames a negative one, so that as the paint flies off the high-revving disc it is attracted to them and penetrates every corner without a droplet being wasted. Undercoats, top coats and colour changes can all be done swiftly and economically in this way. The frames are now ready to join the stream of other components converging on the assembly shops.

We passed complete minor factories engaged in other tasks. There were machines for rolling steel strip into a great coil of wheel rims, cutting and welding them into loops, drilling, spoking and

Brazing the rear drop-outs to the ends of chainstays.

trueing them, and, finally, for fitting the tyres. Other shops were cutting and bending steel tube into many different handlebar shapes, or using giant machine presses to stamp lugs, bottom brackets, brake levers and chainwheels from plain tube and sheet steel. All these parts find their way to the largest automated chrome-plating shop in the UK before eventually moving on, like everything else, to the shop where it all comes together at the end.

A corner of the assembly area. Air-driven tools do up nuts in seconds.

The assembly lines occupy a huge area on two levels with frames, forks and bracket sets being matched together above, before coming down to meet brakes, pedals, chains, wheels, guards, saddles, pumps and all the other parts that go to make a complete machine. Once again we saw how many people are needed to put bikes together on such a scale. One finished bicycle comes off these lines every four seconds, and not the least impressive achievement of the whole business is the coordination required to ensure that it ends up just as specified – and looking like a bike rather than the abstract sculpture in the foyer. Raleigh Industries is a £40 million company and by this time we could well appreciate it. We had walked miles and seen something of the operation, but we could have spent days on the details.

Raleigh keep a constant check on the quality and durability of their machines and do a lot of original research on components and on the products of other companies. Behind the doors of the Product Test Department Mike Davis and his assistants test their own and competitors' bicycles to destruction. This is one of the leading cycle research units in the world, and its expertise is much sought after – and much guarded. Here there are over forty rigs, mostly designed and built by the team itself, for testing tyres, frames, forks, bottom brackets, sprockets, freewheels and alloy parts of every description.

Around us bicycles and wheels shuddered and thumped as they were run on simulated rough roads or on and off kerbs; a quiet little rig in another corner relentlessly twisted and flexed a selection of alloy handlebars and stems up and down, up and down. Near by, a child's pavement cycle was spinning away merrily with weights fixed where the pedals should be. It would not stop until the bottom bracket broke up. We were impressed to find that Raleigh's standard test for all their adult bikes is 6,500 miles without failure at 25 mph on simulated Belgian *pavé*. The Motor Industry Research Association (MIRA) reckons that a car does well to last 1,000 miles on this notorious surface. Of course, this laboratory work is also supported by research on the road using cycles fitted with strain gauges which record their findings on paper in a back-pack carried by the rider. Many of the Department's findings are valuable to industry as a whole, and even Campagnolo of Italy altered the design of their titanium bottom-bracket axles following a report from Nottingham.

If international safety standards for bicycles are ever to become possible and credible, then the experience of research units like this should play a big part. So far, alas, many of the recommendations of the American Consumer Product Safety Commission on bicycles seem to have been drawn up by non-cyclists with sometimes absurd ideas of what is 'safe' and what is legally enforceable. For example, the demand for little cages to stop rear derailleurs from 'protruding' (the cages protrude instead) seems merely trivial. On the other hand, safety standards on alloy castings would be a good thing, for seat pins and cranks still break, even if fairly rarely.

The work of Mike Davis and his team is also of great use to the Raleigh Specialist Bicycle Development Section at Ilkeston on the outskirts of Nottingham. This was the last assignment in our visit and we had been looking forward to it.

Ilkeston is managed by the large and ebullient Gerald

A child's toy trike runs under load on a bumpy roller 'road'.

O'Donovan, who was head of Carlton Cycles prior to their merger with Raleigh. The Carlton name goes back to 1896 at the Old Forge, Carlton, and O'Donovan's father took control in 1939. Now, with some twelve highly skilled workmen, Ilkeston produces competition framesets for the TI–Raleigh pro team in Europe, for Mid–Let–BSA in Britain and for private customers. Gerald O'Donovan's experience goes back to frame-building at Carlton and the days of its successful racing team in the sixties, and he has many contacts in Europe, so his views on development and construction carry considerable authority. He and George Shaw, manager of Carlton–Weinmann, gave us some idea of the problems of design and also of those facing a team manager who must cope with many different preferences, physical types and psychologies. Some pro riders need to be 'set up' mentally for a specific race almost as carefully as their machines.

Since its establishment in 1974 the small Ilkeston plant – no

Gerald O'Donovan (on the right) discusses steering geometry with team mechanic Jan le Grand and young Dietrich Thurau, a rising star then riding for T I–Raleigh in Europe.

bigger than an average commercial garage – has become one of the most advanced lightweight-makers in the world. They build team bikes in 531, and they have pioneered the use of Reynolds 753 tubing. When we visited them they were still the only builders in Britain to handle this demanding material. 'Seven-five-three' is so extremely strong that very thin double-butted tubes can be used with perfect safety. Several gauges are available for different purposes, from super-light tube for short time-trial and pursuit work to a heavier set for six-day track racing. Whatever the purpose of the frame, Ilkeston reckons that 753 will save about one and a half pounds over conventional 531. This makes it comparable to work in aluminium and titanium, but with all the advantage of steel's superior stiffness. But it must be brazed with a special amalgam at extremely low temperatures if it is to maintain its strength and it must be built to absolute accuracy first time as it does not take kindly to second thoughts and resists all attempts to 'cold set' it into alignment afterwards.

We watched a 753 frame being worked on with a hand-held

A 753 track frame is tested at Ilkeston. The bottom bracket is pulled out of line by over 180 pounds' pressure, while the rest of the frame is held rigid at head tube and rear axle. This is one of the frames made for Roy Schuiten's attempt on the Hour record.

oxyacetylene torch and there was no sign at all of the metal glowing. The bullion brazing rod is held in the flame against the edge of the hot lug, and as it melts it is drawn under by capillary action to make the join. Gerald O'Donovan explained that the older style hearth-brazing plays a gas–air flame over a larger area of the lug and tubes. This 'softer' flame guarantees a certain minimum strength to the join and it cannot damage the frame seriously. However, it cannot equal the results possible with careful use of the smaller and more controllable oxyacetylene flame. But time and skilled hands are needed. Whatever kind of tubing they use, and whatever brazing method, frame-builders all know that great care must be taken to localize and control the flame so as to avoid the weakening that too much heat can cause by making the brass or bronze enter the crystalline structure of the steel. Double butting and the right kind of steel alloy also help to resist this effect, of course.

We gained an insight into Ilkeston's attention to design detail when they showed us how they calculate a racing bike's steering geometry by taking into account even the section of the tubular tyre that will most often be used on the front wheel. They have a selection of 'slave' wheels already fitted with different tyres for this purpose. We watched a test rig apply fearsome twisting forces to the frames – which must not be deformed by more than two thousandths of an inch – and also saw a roller road which hammers loaded bikes at 30 mph over simulated cobbles for days on end. We saw the frame on which Dutchman Hennie Kuiper won the 1975 World Championship for Raleigh. He had become so attached to it that Ilkeston agreed to repair it after a crash, although this is against their usual policy with 753 frames. Visiting other specialist lightweight-builders we were to see creations which might have been more finely and elaborately finished than these Ilkeston cycles, but there can be very few better-engineered racing machines anywhere in the world. We had found ourselves talking to men who have a passionate enthusiasm for bikes and bike racing for their own sake.

DAWES

The next stage in our voyage round the cycle industry took us back to Birmingham again, to the Dawes factory at Tyseley – not very far from the Reynolds works where we began. Just across the road, indeed, we could see yet another well-known name in the cycle trade: Renolds, who make roller chains of every description.

The Dawes building was smaller than we expected – the office building alone at Raleigh would have housed it – and yet inside it was laid out in a spacious and airy open plan. Dawes prided themselves on being a family firm and their roots in Birmingham industry go back to 1904, when Richard Dawes's grandfather and his partner produced the Humphries and Dawes OK Supreme motorcycle. The bicycle company was founded in 1926.

Half a century later, Dawes was the largest independent cycle-manufacturer left in Britain, achieving an impressive annual output with a workforce of only about 130, including office staff. By comparison, Raleigh began in 1888 with 200 people. Dawes emphasized that their methods struck a balance between quantity production and individual building in the old style. At first the Company specialized in light machines for the clubman. Only later did they come to make heavier roadsters, just as in the sixties they added small-wheelers to their choice of sports models.

We could not have known it when we visited the factory, but within a year and a half Dawes were to lose their independent status. Problems and costs in distribution and marketing had finally called for greater financial resources than a company of their size could command. They could not match the economies of scale and the diversification so evident at Nottingham.

A medium-sized firm like Dawes builds its frames by hand on an assembly-line principle. About thirty men work in the building shop and each frame slowly takes shape as it is passed between them from bench to bench. First the tubes are selected from the racks – they can be plain gauge from Tube Investments or 531 from Reynolds; then they are cut and pinned to the correct angles on a special table while other men set up the forks. The tubes are brazed by one or two workmen with a gas–air torch on an open brick hearth. Finally the lugs are filed smooth and the completed work shot-blasted and bonderized ready for rust-proofing and stove-enamelling. The spraying is also done individually by a couple of men with guns and protective clothing. The frames are even moved by hand to stand in neat lines of a dozen or so waiting for

Cycle tubing is checked in the Stores.

the next stage, and we spent a lot of time being careful not to knock them over like rows of dominoes. At each stage we found that the speed and the 'eye' of skilled individuals could produce work that would take vastly more elaborate and expensive automatic machinery to match. But of course not nearly so many units can be produced at the end of the day.

The next step is for the completed frames, enamelled, with fancy lug-lining and transfers, to go to the assembly benches where the components are added to make complete bicycles. Off to one side the wheel-builders have a room to themselves where they sit, secluded like crofters at their spinning wheels, lacing, tuning, spinning, tuning, heads cocked, attentive to the flashing spokes. There are jigs with micrometers to help with this, but experienced workers can true a wheel almost wholly by eye and touch.

Machines are produced in short runs, working on one model for a short period before changing over. While we were there they were *building* Kingpin frames, *enamelling* sports frames, *assembling* children's bikes and *checking and packing* Kingpins again. Thus the system has variety built into it and many of the workers can exchange jobs with each other. Again, at the assembly stage each

A Kingpin frame being brazed with a hand torch.

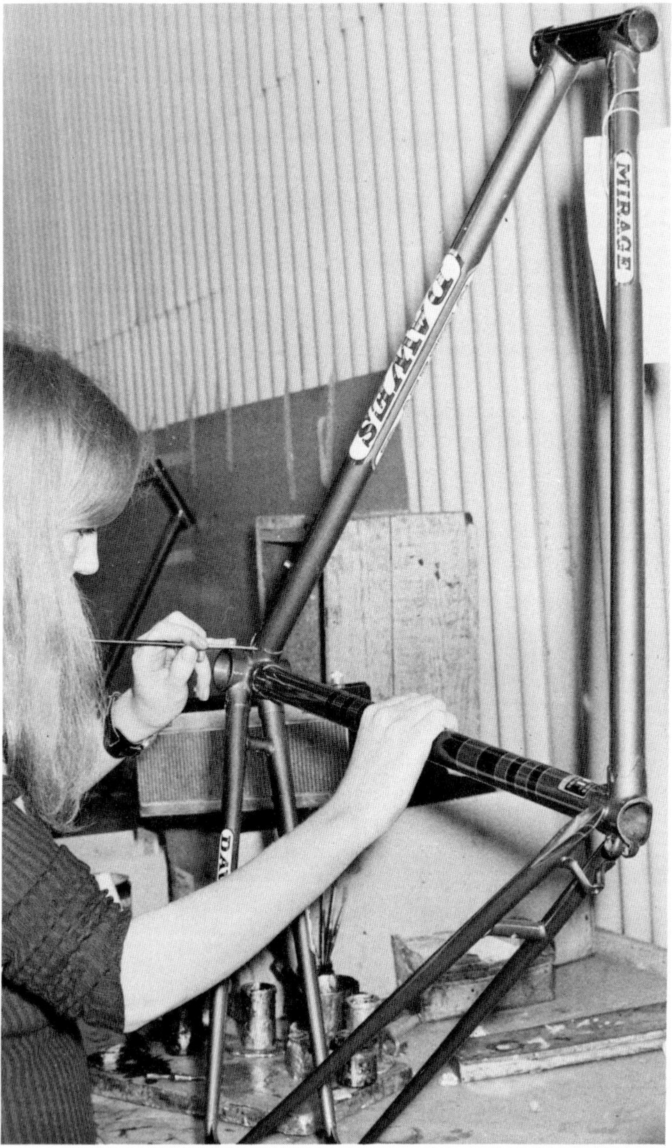

Lug-lining a 531 plain-gauge frame.

person puts together a complete cycle from the box of parts at his bench. We met a man who could set up a bottom bracket in two minutes, complete with cranks, chainwheel and cotter pins. He made it look so easy. All cyclists should have him in their saddlebags: much better than those little dumb-bell spanners. With so few power tools in the building it was relatively quiet, and there was a lot of good chat going on as well.

These factors, and the small size of the building with its relatively open lines, did seem to create the feeling for everyone that they were involved in something worthwhile from first to last. The Carlton factory at Worksop produces its machines along similar lines, a single man being responsible for a completed bicycle. They said that it results in better quality control as well, but older traditions, union opinion and the sheer size of the Raleigh plant make this approach much less practicable at Nottingham.

It was in the Dawes stores building across the yard, however, that we gained our most interesting insight into the complex nature of business for the smaller manufacturer and the industry at large. The components in those bins and racks had come from many sources and many countries. Most cycles today are international machines. (Surprisingly, perhaps, this has always been the case – in the 1890s, for example, American machines with English parts were regularly marketed in France.) Thus, in the Dawes stores the wheelrims might be British (Metal Pressings or Birmalux) or Swiss (Weinmann) or Dutch (Schothorst); the spokes might come from Birmingham, the hubs from France (Pelissier) and the Michelin tyres from Stoke-on-Trent, while the mudguards could be German (Esge) or Bluemels from Coventry. We saw alloy cycle pumps from France via Coventry (AFA) and celluloid ones from Northampton. Steel handlebars come from Raleigh, as do Brooks leather saddles, and also British were alloy bars and stems (GB), and chainwheels and bearing sets (Nicklin, T. D. Cross). Derailleur and hub gears, on the other hand, arrive from France and Japan (Huret, Simplex, Shimano). In some cases Weinmann brake cables from Switzerland are married to outer sleeves from Birmingham (S. J. Clark).

Ordering, accounting and ensuring the supply of components from so many scattered sources is a major job for any manufacturer and constantly changing international currency rates do not make it any easier. The detailed specification of machines is always changing

The angles on a special racing frame are set up and checked with this American instrument.

in response to the ups and downs of the market, and the dangers were explained to us, from Dawes's point of view, of being too dependent on any single source. They regretted the loss of competition, particularly from British products now only memories. This particular litany includes the Airlite hub, Chater Lea pedals and cranksets from Letchworth and the splined cotterless chainset from Edward Williams of Birmingham. At almost any club meeting there will be a veteran still running on some of these parts and chanting their praises to young riders who pray only to Campagnolo.

We asked Dawes about the current growth of interest in cycling and about the commitment they began to make to the sport in 1975. In that year they signed on two professional racing men: Trevor Bull, National Sprint Champion in 1975, and National Pursuit Champion Phil Bayton, co-sponsored by Shimano Dura-Ace equipment. Although these men are professionals, like almost all British riders they have to take other jobs as well – Trevor Bull was working for Dawes in customer sales. This is because the relative lack of public interest limits the financial support that a company can give. On the Continent there are over thirty teams and many hundreds of thousands of pounds a year are spent on professional cycling, but there are not nearly so many sponsored pros in the UK and the number of teams varies from year to year, as do the riders within them. In 1978 Bull moved back to Carlton, while Bayton joined the large Holdsworth–Campagnolo squad.

Working for Dawes, Bayton and Bull often found themselves in somewhat unequal competition, but at the same time this increased their publicity value and this is what pro cycling is about. Bayton, nicknamed 'The Engine', is famous for his aggressive attacking style. He once explained that his job is to get his sponsor's name into the news, and if he cannot win a race he will still break away early and make the others work very hard to catch him. Riding flat out like this he often wins primes and lap prizes along the way and so again he earns publicity (and more money). Looking at photographs of him crouched on the bike with his face and jaw contorted in effort, we reckoned that he more than earned his keep as a cashman. Dawes were following this favourable publicity with a limited production of out-and-out racing frames as ridden by Bayton and Bull in their cerise and blue livery.

The racing market is a small one, however, and it was competition from a mail-order folding cycle that Dawes were discussing when we visited them. They had bought a model and found it to be reasonably good quality for its low price. The machine was imported from Yugoslavia, so the savings were made in labour and

Phil Bayton at work for Dawes–Dura-Ace.

subsidies. We gained some insight into the tight margins that companies have to operate within and the worries that go with them. Once again we saw that cycle manufacture is labour-intensive, and that the price of this manual work cannot be escaped. Cost-cutting is difficult, since safety is always a primary responsibility and there is so little on a bicycle that can be spared. What can you do? Use cheaper transfers? The ghost that haunts cycle makers is a cheery voice that says, 'Seventy *pounds* for a *bicycle*? Why, I bought my old one for only twenty, and I think it's still in the shed. I'll just get it out and clean it up.' The fact that everything else in life is more than three times as expensive does not seem to occur to the speaker, for he is a democratic shade and he believes that everyone should pedal for ever at 1950s prices.

We raised this particular ghost again when we discussed the astonishing durability of the bicycle. A recent Italian report claimed that their industry was in financial straits because, at fifteen years, the average life of a bicycle was just too long. They reckoned that only three years less would maintain sales at a better level – better from their point of view, of course, but not necessarily for the peasant carrying a piglet on his cross-bar. In Oxford and Cambridge, where so many of the University students cycle, machines seem to last for ever and giant upright gas-piped antiques, complete with wicker nose-baskets, change hands very cheaply from generation to generation. Perhaps it is this long tradition (and a certain genealogical similarity) that gives them their affinity for the black iron railings outside tea-shops.

From this point of view the real significance of the small-wheelers that appeared in the mid sixties was commercial as much as technical. In other words, they stimulated a new interest and boosted sales at a time when the industry badly needed it. The same has been said of the Raleigh Chopper. Whatever its drawbacks in the eyes of cycling purists, this machine galvanized the British youth market in 1970 and created a new demand for a kind of cycle (and a kind of cycling) that had scarcely existed before. Raleigh first marketed this machine in America and still sell thousands of them there, while more than three quarters of a million Choppers are on the road in the UK alone. The new Raleigh Grifter, with (in our opinion, at least) more logical lines and a gear-changer safely on the handlebars, looks set for a similar success. One of the Nottingham company's strengths has always been the financial security provided by its toy and children's cycle sales.

When we had found our way out of the tangle of factories in the Midlands, passing familiar brand names at every turn (from spectacle frames to truck air-brakes), our last task was to take a closer look at the exclusive and expensive machines provided by the specialist builders. So we left Birmingham and turned south on the motorway towards Worcester and Dave Moulton.

FRAME-BUILDERS

The specialist frame-builders live in a different world from the large manufacturers. There are scores of them in the UK, scattered across the country in back streets, small shops, old garages and even a converted chapel. On these sometimes unlikely premises skilled individuals will hand-build a complete frame to a customer's specification. They use the best materials and their respective merits are discussed and debated by enthusiasts with an eye for detail. At its most typical this is essentially a cottage industry (sometimes literally) and orders may take anything from six weeks to six months or more because these firms are just not geared to quantity production.

A large firm like Holdsworth, however, produces complete sports bicycles as well as top-line racing frames. They also market Claud Butler and Freddie Grubb models, and their racing-team manager Roy Thame distributes lightweights through the Company under his own name. The bigger frame-builders like Witcomb, Bob Jackson and Mercian also manage to supply off-the-peg framesets to dealers and to export their work as well. The scene is a complicated one, however, and in some other cases frames are made up by various small builders to be sold complete or 'in the rough' to dealers who will finish them or deck them out as they choose. The name on the down tube is not invariably an indication of where it was built, much less who built it.

Most frame-makers are very small businesses. Some are long established, like Harry Quinn who have been in Liverpool since 1901, or Ellis Briggs of Shipley in Yorkshire, or Rattrays of Glasgow who have been making their Flying Scot frames since 1925. Publicity-conscious firms like Ken Bird in Kent will sponsor

Pearson's cycle workshop in Sutton at the turn of the century. Founded as a blacksmith's in 1860, it began to make Ordinaries in 1885. The family is still in business at the same address.

cycle clubs so that time triallists and racers will win on their machines or with their names on their jerseys. (Amateur racing rules allow for this.) Other businesses rely almost entirely on word-of-mouth local support, and some may seem reluctant to build you a bike even when you ask them. A catalogue of frame-builders reads like some strange Doomsday Book. Slender, responsive machines, widely coveted in Europe and America, display surprisingly down-to-earth surnames. Such a list would include E. G. Bates, R. J. Quinn, Don Farrell, Reg Barnett, Alec Bird, Tom Crowther, Jack Hearne, F. W. Evans, MKM, Dave Russell, Fred Baker, Jack Taylor, Ken Lazell, Norman Fay, Shorter, Higgins, Hetchins, Roberts, Walvale and Woodrup – there are more. They are all music to the ears, and they make the car industry's flamboyant labels – 'Mustang–Cougar–Jaguar–Tiger GT' – seem more than a little adolescent by comparison.

The number of builders in America is growing every year and names like Don Millberger, Art Stump, Masi, Tom Ritchey and

John Patston and 'stoker' John Woodburn on a record-breaking ride. Their short-wheelbase racing tandem weighs only 31 pounds complete and in 1976 it cost nearly £800.

Tanguy Cycles are becoming established. Albert Eisentraut of California runs frame-building classes, Colin Laing in Tucson emigrated from England, where he learned his skills, and Proteus Design is a forward-looking group from Maryland who even market a do-it-yourself building kit complete with instruction book and decals. There are scores of prestige builders in Europe and among the best known are the romantically appealing names of André Bertin, Alex Singer, René Herse, Jacques Anquetil, Mercier, Lejeune and the larger Peugeot, Motobecane and Gitane in France; Cinelli, Benotto, Viner, Bianchi, De Rosa, Pogliaghi and Colnago in Italy; Zeus in Spain, and Flandria and Superia of Belgium. Impressive lightweights are produced in Japan by the larger manufacturers, Bridgestone, Fuji, Miyata and National Panasonic. Japan's high-quality components industry is already an important force in the international market and there can scarcely be a rider in the West who has not seen or used equipment from Shimano, Sun Tour, Mikishima, Sakae, Sugino or Dia Compe.

To come closer to home, however, the essence of the traditional builder's craft is still to be found in small workshops where kettles stand by the brazing torches, where orders worth hundreds of pounds are kept on an old spike and a week's output can hang from four hooks on the ceiling. There is great satisfaction in meeting the builder at his bench, when planning your own machine down to the last detail.

At last we left the motorway at Worcester and drove through pleasant country towards Deblins Green. Dave Moulton used to work as a machine-tool fitter in the aircraft industry and he began building bikes for himself and his friends in his spare time. He only went into full-time production in January 1975, but already he is known as a man with outspoken ideas of his own on cycle design. For these reasons we thought he would be an interesting man to talk to, and we were right. (He is, incidentally, no relation to Alex Moulton, the engineer who pioneered the small-wheeler and the hydrolastic suspension for British Leyland's Mini.)

Hill View Works is a large black Nissen hut in the countryside which Moulton shares with a local motor mechanic. Here, in a tiny shop partitioned off at the back, he works with Andy Thompson. They build about three or four frames each week. Simply preparing the materials takes a lot of time, as tubes have to be cut and mitred and all the lugs and drop-outs have to be chamfered and cleaned up by hand before they can be used.

Moulton showed us how his frames are gradually put together, beginning with the flat steel table where angles and lengths are marked out and the tubes prepared with flux and 'tacked' together with small blobs of brazing to hold them in position. When this is done the tubes are taken to another jig and final brazing can begin. A more traditional method of holding the tubes is to drill them and pin them but Moulton prefers not to pierce them in any way. Like the men at Ilkeston, he uses a hand-held oxyacetylene torch. He heats the lug and its attached tubes until they glow hot at about 820 degrees centigrade. Then the brazing is done with a thin rod of copper–zinc alloy, or silicon bronze. Silver solder can be used for 531SL and Columbus super-light tubes which have to be worked at

Ready for building – lugs filed, cut out and polished.

even lower heats, but it is an expensive material and adds appreciably to the overall cost.

In terms of engineering sophistication frame-building is not much more than super-accurate plumbing, but the skill comes in handling the materials, controlling the heat, judging when the molten brass has penetrated the entire joint and being able to estimate and minimize the effects of expansion and contraction. The lugs are therefore brazed together in a certain correct order, always allowing for proper cooling before moving on to the next stage. Dave Moulton has made three jigs for checking and ensuring the alignment of the main frame as he works on it, and he uses another two for the rear triangle and front forks. Once more it comes down to skill, however, for jigs alone are no guarantee of a perfect frame and an inexpert worker can actually build stresses into the tubes by using his clamps carelessly. When a frame is completed it is sandblasted to remove the slag left by the brazing flux and Moulton then spends about four hours hand filing it to remove excess brass and to feather the points and edges of each lug down to about half their original thickness. This lengthy attention to detail from start to finish is

Dave Moulton at work with torch and brazing rod.

common to all the best specialist builders, and although some of their refinements may not be strictly essential they are still very pleasing to owner and builder alike. Moulton himself goes for very functional clean lines.

Even now there are still several extras to be added according to the customer's specifications: a button, perhaps, to stop the gear levers slipping on the down tube, or integral bosses for them; special brake bridges, bottle-cage bosses, pump pegs and cable guides and tunnels. All these have to be attached very carefully and for this some builders use silver solder. The next stage is priming, in which a mild phosphoric acid is used to provide a good key for the coats of paint which follow. The transfers are added, lug lining painted in and then, finally, a clear coat of lacquer goes over all. It seems to be difficult to find really good finishes these days. Much of the success of stove enamelling lies in temperature control and older firms like to keep these details secret. Modern chrome plating leaves a lot to be desired as well, compared to pre-war quality, and some builders eschew it altogether. If the customer has asked for it the builder will also fit the headset and bottom bracket to the frame, and this is a good idea as he should have the special tools which make the job sure and easy.

Dave Moulton does not support the current craze for extremely short wheelbases with ultra-steep head angles. He maintains that proper design can give a bike lively and fast cornering qualities without the penalties of 'twitchy' handling and pedals that overlap the front-wheel. Some time-trial machines are almost unrideable except in a straight line. In America even tourists will ride on frames that are really more appropriate for short circuit races. The long miles must take a terrible toll on their backsides and their wrists; this seems a big price to pay for fashionable lines. On the other hand, of course, improved road surfaces do mean that today we can use 'stiffer' bikes, and tourists in the seventies are comfortable on designs that would have been too extreme even for early Tour de France heroes like Eugène Christophe.

We agreed that it was surprising to find that many top Continental pros know little of the design details of their machines. The *coureur* often leaves such matters to his builder. In Britain and especially in the United States, however, club riders can spend hours happily arguing over the finer points. Cyclists are great traditionalists, and yet at the same time the thought of the 'perfect' machine haunts them like some holy grail. So fashions come and go.

Every large club has its equipment freak, the man with a fortune in parts in his garage and what British riders call cigarette-paper clearances between his rear wheel and his frame. (We know a rider whose back tyre regularly wore away the paint on his seat tube.) Such owners and some builders will go to extraordinary lengths to save a few grams' weight on their machines. Sometimes this clearly goes beyond all practical point and becomes a mental event of some sort rather than a question of physical advantage. Indeed, heavily drilled, cut-out and slotted frames have to be treated very gently, and even then the wind resistance of all those holes may amount to more than is gained by loss of weight. Why not slim and lose a whole pound yourself? Nevertheless, lightweight design at this level can be a fascinating blend of engineering necessity, the builder's technical virtuosity and the customer's preferences. There is no denying the attraction of the beautiful finish on the lugs of a Roberts frame, or of the fine engraving and drilling done by the Italian Viner workmen on Campagnolo and Cinelli equipment, or of the attention to detail that Moulton gives to his drop-outs and brake bosses. A good hand-built frame is a beautifully finished machine suited in every detail to its owner's tastes and proportions. But it must be said that in some cases, by its very nature, hand work can be fallible, and in the course of our travels we did hear sad tales of misaligned frames and weak joints. It seems to be something that happens now and then, but most reputable builders will treat such complaints seriously and do all they can to put matters right.

Dave Moulton is an articulate man with a rather quiet and down-to-earth manner. He approached the market with an engineer's eye for the most essential issues in both design and selling. He recognized from the outset that he had to break into a small and specialist trade and he did this by skilful publicity and by assuring his customers that a lot of his own time and skill goes into their machines. To follow this up he managed to interest some top riders in his work. He has written on the subject of cycle geometry and design and questioned some traditional assumptions, learning from his mistakes and his successes alike. Obviously he cannot make many frames, but on the other hand he knows that they gain a certain rarity value among cyclists who are still looking for the holy grail and are prepared to spend a bit more on the search. When we visited him he already had orders from Canada and Denmark as well as his British trade. He knows that exclusiveness is itself a valuable commodity in a small market of enthusiasts.

Of course, when all is said and done about the niceties of design, decoration and construction, these features do not win races. Perhaps we should end by letting Frank Dickens point the moral from *The Great Boffo*, his delightful book for children. Our hero punctures on the last leg of a race but manages to win it by leaping onto a

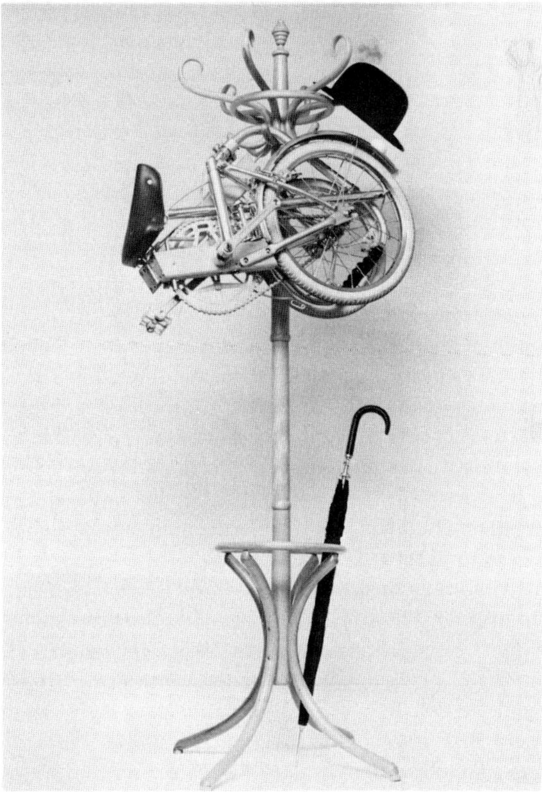

grocer boy's delivery bicycle and pedalling off at a great pace with basket and bottles rattling at the front. At the end of the story a grateful Boffo gives the lad a sleek new racer, telling him,'. . . but remember my words: the machine is not as important as the man.' It is well worth saying. After all, more than one rider has completed a stage of the Tour de France on a machine borrowed from a spectator.

In our brief visits to the industry we learned a lot, but by no means all, about the varied and endlessly detailed world of bicycle design, manufacture and selling. We were treated to hospitality, fascinating discussions and many different and sometimes contradictory points of view. This is how we saw it, or most of it.

6 Cycle Sport:

The Clock, the Track and the Mountains

The world of modern cycle sport is as complex as a clock – wheels within wheels, indeed. Its true home is with the professional riders of Europe, particularly in those most powerful cycling countries of Belgium, France, Italy and Holland. Professional racing is a hard life. A rider will drive hundreds of miles across Europe with bikes and wheels in the back of his car. He will travel from fair-day *kermesses* on the cobbled streets of small towns in Belgium to six-day events on the indoor velodromes of Italy. He cannot always win, of course, but there are still special classifications, *primes* and 'hot spot' awards to be gained and the prizes may go to thirtieth place. An average *domestique* – a rider whose job it is to support and protect the stars on the road – can expect a basic annual contract of around £5,000 from his team's sponsor, plus bonuses if he wins or places highly. On the big races top winnings are also shared out among the team. If the rider is a star his annual contract will be for a very considerable sum and he will also collect a lot of appearance money for drawing the crowds to the event.

For enthusiastic followers of the sport the details of performance, equipment and trade support, the consideration of the riders' strengths and weaknesses and the spotting of tactics, financial deals, sporting alliances and betrayals all provide material for endless café-table debate. The arguments are as complicated as any learned doctrinal wrangle, but the passions aroused may not be so subtle. Indeed, in the early years of the Tour de France riders would literally be in fear of their lives if they trounced a local favourite on his home ground. After the sun set on some of the longer stages, ambushes and sabotage were not unknown. Even today the Italian *tifosi* are a daunting sight as they line an Alpine ascent, scrambling to cheer and help their countrymen and booing their rivals.

Perhaps the most striking proof of the sport's hold on its followers comes from post-war Italy and the story of Gino Bartali's victory in the 1948 Tour de France. While the race was running the political situation in Italy became very unstable and an armed clash between right- and left-wing factions seemed imminent. Italian reporters began to leave the Tour to return home. (The depressed state of the country at this time is graphically depicted in Vittorio de Sica's film *Bicycle Thieves*.) The Prime Minister, De Gasperi, was warned that just a few more gunshots might plunge the whole country into bloodshed. It occurred to him that an Italian victory in the Tour might provide a boost to morale. He telephoned Gino Bartali and asked him to win the race or at least the stage for the next day, 14 July. Gino won the stage.

To win overall was a tougher task, for Bartali had a lot of time to make up and, at thirty-four, his last Tour victory was ten years in his past. Bartali was a hard-headed native of Tuscany, deeply pious and noted for his give-no-quarter riding. He had always been physically strong and so he attacked again and again. Young Louison Bobet, who was later to win three Tours in succession, could not withstand the fierceness of the Italian's riding in the Alps and soon the yellow jersey fell to Bartali. To clinch it for Italy, a member of the team managed to win the last stage into Paris as well. It was one of the great victories: Bartali won seven of the twenty-one stages and helped set a record average speed for the race that was to last for the next six years. All Italy rejoiced. The country had been agog during their hero's courageous assaults in the mountains. The revolution had been overtaken in the Alps. The politicians had cause to be grateful to the Tuscan rider and 'Gino the pious' received the final accolade when he was received by Pope Pius.

RIDING ON THE WIND

The structure of cycle sport is every bit as various as its commercial and political ramifications. At both amateur and professional level it includes many different kinds of competition, from short-distance

The greatest and best-loved post-war rider, the *Campionissimo*, Fausto Coppi. He is out on a training run but the crowds still line the route.

time trialling to stage races over thousands of miles, from cyclo-cross through ditches and muddy slopes to track events on smooth and steeply banked velodromes. Such events demand widely different tactics, equipment and even physical types. Yet, for all its variety, every kind of bike racing has one feature in common – the resistance of the air as you try to cycle through it. Almost everything in the sport, especially the tactics of road and track racing, revolves around the basic physics of air resistance. Let us look at this a little more closely.

At racing speeds of thirty miles an hour on the flat, nearly all a rider's strength goes to overcome the resistance of the air around him. Compare, for example, the world hour records for paced and unpaced riding. The unpaced 'Hour' is the most severe test in cycling, attempted only by the great-hearted. Merckx broke the record in Mexico City in 1972. He depended on magnificent physical and mechanical preparation to carry him through. (His tyres, for example, were inflated with helium.) After sixty minutes of what he later described as 'a constant battle against pain' he had covered 49.431 kilometres, an average speed of 30.69 mph. By comparison the paced hour was set in 1955 by Willy Lohman of Germany at a speed of 58.38 mph behind a special motorbike. The shelter afforded by the pacer speaks for itself in that extra 27 mph.

There has been a growing American interest lately in speed competitions using 'streamliners'. These machines vary from conventional racing cycles with fairings to horizontal bomb-shaped tricycles. From a flying start, speeds of 50 mph have been achieved over 200 metres, but the pace drops off over a longer kilometre course. These 'International Human-Powered Speed Championships' have become an established event at Irwindale in California, but they are very specialized indeed, a kind of pedal drag racing.

Streamlining is not new and its possibilities have been explored from as early as 1913. Two of the best-known exponents in the twenties and thirties were Oscar Egg from Switzerland and the Frenchman Marcel Berthet. In their youth these two riders exchanged the unpaced hour record no fewer than five times. Egg's final record was to stand for nineteen years before it fell to Maurice Richard in 1933. That same year, however, the 47-year-old Berthet came back to ride his 'Velodyne' streamliner for the hour, beating Richard's distance by nearly four kilometres. His modified machine did not qualify for the official record, nor did he expect it to. But he made his point about the effectiveness of streamlining.

At the wilder edge of record-breaking, 'Mile-a-Minute' Murphy achieved notoriety and 60 mph behind a specially adapted railway train in 1899. Also in the United States, in 1972, Alan Abbott was

An Oscar Egg streamliner in action.

paced behind a racing car to 138.7 mph across Bonneville Salt Flats. Such events require giant chainwheels – almost as large as the cycle wheels themselves – and giant nerve, too. Abbott had to be towed up to 80 mph before he could even begin to pedal under his own power.

Such feats are impressive but they can scarcely be called a sport, and, indeed, once or twice is usually enough for even the bravest rider. Nevertheless, beating wind drag is still taken very seriously by conventional racers. On the track riders will wear silk jerseys well tucked into their shorts, and Marcel Berthet even used to tuck in his

Oscar Egg responded to Berthet's success with the Velodyne, and early in 1934 he too came out of retirement to attack the 'Hour'. Here he is training on a recumbent of his own design.

shoelaces and oil his hair flat. Recently one-piece 'catsuits' with long arms and legs and hoods which cover the head have begun to appear. Speed skaters and skiers have already shown the way with their fishskin outfits and although there has been dispute over the catsuits one wonders how long it will be before most track riders look like

Batman. These are still specialized applications, of course, but even the cycle tourist can learn from them and can reduce his drag area by as much as thirty per cent if he avoids loose and flapping clothing. The wind blows on everybody.

THE TIME TRIAL: 'CONTRE LA MONTRE'

In complete contrast to the exotic world of streamliners and the hour record, time trialling is the commonest form of unpaced racing, and in the United Kingdom it is by far the most popular

Sunday-morning time trial in the quiet of a Berkshire village.

branch of amateur cycle sport for both men and women. The task is to ride alone against the clock for a set distance. Courses are registered and accurately measured by the Road Time Trials Council (RTTC) and only officially approved watches and timekeepers are used. This body also organizes the National Championships each year and makes the regulations by which the sport is run. The official courses are over 10, 25, 50 and 100 miles and there are 12- and 24-hour events in which it is the distance covered that is measured and recorded to the last yard. Riders are sent off every minute and there are strict rules against cycling in another competitor's slipstream. The effects of hills and prevailing winds are taken into account by riding 'out and back' with a turn somewhere near the half-way point.

Time trialling is a curiously mathematical sport and the competitor has many searching decisions to take regarding the gear he

Beryl Burton, as streamlined as possible, turns a big gear. The stop-watch on the handlebars will help her judge the ride.

chooses, how fit he is and how fast he can pedal the course. He needs to know himself fully and, if possible, he should know the course as well. The triallist sets himself a target time to meet and he should cross the line without a scrap of energy to spare. This form of effort requires a total concentration which is strangely relaxing once a rhythm has been attained. In team time trials for two to four riders, each man will take a turn at the front to keep the pace up for a hundred yards or so before slipping to the rear and into the shelter of his team mates. Once again, rhythm and concentration are essential – especially since only inches are kept between the front and rear wheels of the bicycles.

Each year the keenest amateur riders in Britain enter their times for the British Best All Rounder competition run by the weekly paper *Cycling*. This averages a cyclist's fastest times for the '50', the

'100' and the 12 hours. The winning average speed is usually over 25 mph. The BBAR for women does not include a 12-hour event and the times are a little slower, except for those of the incomparable Beryl Burton who has won the competition ever since 1959 with an average speed that leaves many men behind. Indeed, in 1967 Mrs Burton became the only woman athlete ever to break a men's record when she achieved an outstanding ride of 277.25 miles in 12 hours.

Some riders specialize in short events like the '25', while others do best on the longer ones, eating and drinking on the move as helpers hand up plastic feeding bottles and sponges from the roadside. It is not unusual for older riders to do well in the endurance trials and some achieve better times than they did as young men. The long rides, and especially the 24-hour events, seem to be something of a British institution. The particular strength of time trialling is that it offers competition for everyone at all levels of ability. Thus there are separate classifications for women, juniors, schoolboys or girls and 'veterans' over forty, as well as a handicap system for all. Each rider will have a personal best time for the distance and will go home happy if he or she manages to improve on it.

When victory is measured to a tenth of a second, it is not surprising that keen 'testers' are always looking for lighter and faster equipment and are haunted by the possibility of a 'float morning', when ideal conditions and a good road surface will allow them to attain even better times. In fact the sport has become very specialized in Britain, and this is what produces the demand for those ultra-light machines with huge chainwheels. It follows that 'fast' courses are equally as much in demand, and one or two have become notorious because of the speeds possible on them. Heavy traffic plays a part in this as well, because overtaking trucks create a 'suction' effect which allows the racer to turn higher gears than he might otherwise manage. Thus, of the top twelve riders in the 1976 BBAR, ten had made their best times for the '50' and '100' events on one particularly busy and well-surfaced 'drag strip' using the A1 in Yorkshire. The implications of this led to heated arguments in cycling clubs around the country. One way to try to offset such highly selective racing would be to have the top contenders compete in the National Championship events each year with everyone riding the same course on the same day. Whatever the controversies, however, it is certain that time trialling will continue to thrive in Britain.

A fellow competitor gives up his back wheel to one of the fast men who has punctured during a '25'. The delay cost a minute.

For many riders cycling would not be the same without the sight of bright jerseys and glittering machines assembled by the road in the early Sunday morning. A colourful caravan of people and cars has suddenly appeared in the middle of the countryside. It is a busy scene. Riders are fixing numbers to their bikes or to their jerseys, rubbing liniment into their legs and inflating tubulars so hard that a wheel rings like a bell if you tap the tyre with your fingers. While some competitors are queueing at the start, the first men off may

At number 120, Alf Engers is favourite for the event and last man off. The brakes, levers, chainwheel and handlebars of his bike have been extensively drilled out.

already be racing for the finish line only a hundred yards away on the far side of the road. They look tired, but recover very quickly to talk of how they caught up with their 'minute man' ahead, or of how rough the road surface was to ride on. Tension mounts for the triallist as the handlers hold his bike upright in front of the timekeeper. If his chief rivals have already finished, the fast 'scratch' man may know exactly what he has to do to win the race. Friends and club mates will also shout time checks at the turn or along the route to help him home. When the thirty-second warning is given, the tester tightens his toestraps and starts to breathe deeply, lost already in his concentration on the task ahead. The handlers take a firmer grip ready for the push-off. It comes like a relentless benediction to each man in turn – a special excitement as the timekeeper intones over his stopwatch: 'Five, four, three, two, one: GO.'

The 'black alpaca' days lasted more than forty years. Frank W. Southall collects a feed during a 100-mile time trial in the thirties. Southall won the first four B A R competitions in a row.

Enthusiasts in the United States are beginning to organize more and more TT events and the US Cycling Federation sanctions courses, just as the RTTC does. The Onondaga CC of Syracuse started a 24-hour event in 1973 and this has become steadily more competitive each time it has been run. Given the difficulty of organizing road races on the highway, it may well be that 'testing' will become as popular in America as it is in Britain.

In Europe it is road racing which holds sway and the time trial is much less prominent. One reason for this is that the British authorities banned massed start cycle races on the public roads as early as 1888, and time trialling was introduced as a 'safe' form of competition. Thus at the turn of the century British triallists would meet secretly like duellists at dawn wearing anonymous black tights and alpaca jackets intended to make them inconspicuous. They could not even publish the details and results of these meetings. During the same years whole villages on the Continent would turn out to watch the great publicity-conscious road races as they stormed across the countryside. It was not entirely a question of unsympathetic British officials, for it seems likely that road racing also thrived in France because of that country's more dispersed and rural population. In the larger cities and population centres of Britain roads were always more crowded, and it was football which early caught the imagination of the urban masses.

Whatever the reasons, the time trial in Europe has developed differently. It is usually held over shorter distances and hillier ground than is common in Britain, or it features as a very demanding part of a stage race. The most famous European trials are the *Grand Prix des Nations*, held for both professional and amateur riders over sixty miles or so outside Paris, and the Italian Baracchi Trophy, which is a two-up team trial for a similar distance. These are distinguished events in the sporting calendar, taking place in October at the end of the road-racing season.

British triallists do not do as well as might be expected when they cross the Channel, despite their specialization in this form of riding. This is partly because they tend to concentrate on long distances on flat courses and, indeed, it is often suggested that such specialization spoils riders for the tougher all-round competition of road racing. Road men need to be more versatile to match the changes of pace

Time trialling as a professional. The following car provides encouragement, advice and even a spare bike if necessary. Jacques Anquetil against the clock in a stage of the 1962 Tour, a race he won at record speed.

and terrain that a race involves, and on the whole this shows to advantage against the clock as well. Alf Engers, for example, has been British amateur '25' champion for six consecutive years, with a record for the distance of 51 minutes exactly: an average of 29.412 mph. The top Continentals can time trial nearly as fast on much hillier courses, though of course they are professionals: witness Freddy Maertens's superb 29.3 mph over the tough 23-mile Le Touquet stage on one of the hottest days of the 1976 Tour de France. Yet the top triallists still excel at judging their pace to the second and at forging a smooth and powerful style. Jacques Anquetil was a master of these skills and they contributed to his victories even in the major stage races. There are few great road racers who are not also good against the clock.

HILL CLIMB AND CYCLO-CROSS

The hill climb is a sudden-death version of time trialling, a race against time over a course seldom more than two miles long and sometimes as short as four hundred yards. In Britain these events are usually run in October towards the end of the racing year, and the 'anti-gravity' men make this their speciality. Light and wiry riders are best suited for speed on the slopes and their machines are similarly stripped to the minimum. Thus the climber often rides a fixed wheel with only a front brake, his cranks may be longer than usual to give extra leverage and he will need the security of double toe straps to keep his feet in place as he pulls on the pedals. The National Hill Climb Championships are entertaining events with an end-of-season Autumn carnival spirit to them. Hundreds of bikes are stacked along the winter hedgerows, as bright and warmly clad cyclists line the narrow route to cheer the racers on in their heart-bursting efforts to get to the top in the shortest time.

Cyclo-cross began as an off-season and off-the-road activity for racing men, but it has developed into a specialized winter sport in its own right. The courses are usually circuits of about two miles a lap and the going can vary from fast riding on grass slopes and tarmac to an exhausting slog carrying the bike over mud, streams, ditches, logs and hurdles. There are both professional and amateur categories in

’cross, and in the U K, at least, they often ride together. The evolved cyclo-cross bike has a longer wheelbase, a lower saddle, lower gears and shallower angles than a road racer, and it will have extra clearances at the wheels and brazed-on cantilever brakes to avoid mud clogging the works. This is important because special knobbly tubulars are used to give grip and they throw up a lot of dirt. Every rider has his own favourite modifications in this sport and some have several bikes so that helpers can wash the mud off as they change machines on each lap. Unlike other forms of cycle racing, in ’cross the rider wants to lead from the front, where he can stay clear of crowding at the obstacles and other men’s crashes. It is a hard and exhausting sport demanding the ultimate in agility, bike-handling and stamina, but tactically it is relatively simple – plug your way to the front and stay there. Some riders will let an opponent almost catch them, only to draw away every time he gets close – again and again. It is a dispiriting ruse.

HARD RIDING

Hard riding is not a sport in the strict sense, but it does deserve a place in any account of what cycling has to offer. France and Italy provide great excitement for the fast tourist in *cyclo-sportif* and *randonneur* events. One of the most famous of these is the Paris–

Cycle speedway is a growing sport in the UK, with over two thousand participants each weekend. Racing is done on oval tracks, with groups of four riders on specially built bikes.

Brest–Paris, a marathon 750-mile rally which has to be completed within ninety hours: 200 miles a day for three and three quarter days. Competitors must fit their machines with lights and mudguards and get their *carnets* stamped at each control town. Hundreds of amateur riders turn out for the P–B–P, which is organized every five years. This tourist marathon actually began as a road race in 1891, when *Le Petit Journal* of Paris promoted it to help increase sales. The famous professional Charles Terront won at an average of nearly eleven miles an hour despite a succession of punctures to his balloon tyres. The roads he covered were little more comfortable than modern farm tracks and Terront was riding a machine weighing forty pounds, complete with a fully enclosed chaincase. The race was considered to be so demanding that it was organized only every ten years. Its present popularity says a lot about the improvements made in roads and equipment.

British club riders on a hard training run in February.

The Flèche Velocio takes place every Easter in memory of 'Velocio' – Paul de Vivie, a leading exponent of cycle touring in France. This event has a 40-hour section for the more relaxed tourist and a 24-hour *flèche* for the fast groups of riders who aim to cover some 300 miles or more. These are not road races or time trials in the strict sense and they are not open to riders who hold racing licences. Nevertheless, the *touristes sportifs* wear near racing gear and are in every way serious and fast *coureurs*. Vivid accounts of the colour and drama of French club cycling can be found in J. B. Wadley's book *My Nineteenth Tour de France*.

By comparison the British Cyclists' Touring Club offers rather less excitement, although there are many hard riders within it and some strenuous tours are organized. The annual Cycle Tourist of the Year competition calls for all-round ability and a very British blend of map-reading and observation skills. The CTC also has a Best All Rounder Time Trial Competition. 'Reliability trials' are often undertaken by British club riders in the early months of the year. Different times are allowed for groups of different ability over distances of fifty to a hundred miles and some of the 'scratch' groups move at near racing speeds. *L'Ordre des Cols Durs* was founded in France but it has a British section for cyclists who amass more than 20,000 metres of pass-climbing in the year.

Some prefer to travel using green roads, byways and tracks.

American hard riding is not yet as organized as *le cyclotourisme* but it shows great potential. Thus the two-day Tour of the Scioto River Valley (TOSRV) attracts as many as 3,000 people each year. The old and the young ride on every kind of machine, while the fit men set out to cover the two 100-mile stages as fast as possible. The Bicentennial in 1976 saw the founding of a 4,250-mile coast-to-coast 'Bikecentennial' route. Cyclists can ride alone, or in groups with experienced riders in charge. Each stage of the journey has nominated stop-over points and a wealth of information is available about repair shops, touring hints and the countryside along the way. The scale of the project and its concern to involve people of all abilities is impressive, and it is to be hoped that the Bikecentennial route prospers for tourists and hard riders alike.

The most romantic distance test of them all is the 870 miles through the length of Britain from Land's End to John o'Groats. The 'End-to-End' goes back to 1885, when it was covered in just over 6½ days on an Ordinary. Seen here is one of the most famous of these rides, that by Crimes and Arnold on a tandem trike in 1954. They broke four records in the process and finished in 52 hours 26 minutes. The current solo record is 47 hours, 46 minutes and 35 seconds, set up when R. E. F. Poole 'broke' two days in 1965. All such events are timed and monitored by the Road Record Association.

TRACK RACING: ON THE BOARDS

Track racing takes us back to one of the oldest and most specialized forms of cycle sport. Tracks vary from broad open-air circuits on concrete, or even grass, to the tight and steeply banked hardwood ovals of the indoor velodrome. On a 250-metre track the banking at each end of the two short straights can reach fifty degrees, and riders use this height by climbing right up to the spectators at the railing before swooping 'downhill' into the attack. Indoor six-day tracks can be very tight and frighteningly steep, but the official championship distance is less extreme at 333.33 metres to the lap.

To cope with sudden out-of-the-saddle accelerations and with the stresses imposed by the gradients, sprint bikes are built to be extra strong and rigid, with a high bottom bracket so that pedals do not touch the banking. The 'track iron' has no brakes and a single fixed rear sprocket means that the rider is completely at one with the rear wheel when he straps his feet into the toe-clips. The typical sprint gear is about 92 inches, which could be the top gear on an average touring bike. In order to transmit the power from the pedals the chain might have an inch pitch, which is held to be much stiffer than the usual half-inch pattern, or it might be the new ten-millimetre Shimano system, which uses a smaller and therefore lighter chainwheel and sprocket. Australian John Nicholson, World Pro Champion in 1975 and 1976, uses this set-up and also does without socks, toe-clips and straps, since his shoes are bolted permanently to the pedal spindles. This gives great security but it means that riders have to have their bikes laced onto them each time they ride.

American sprinter Sue Novara on the boards. She has twice been Women's World Champion.

Track frames are extremely close-coupled, with upright angles and tiny wheel clearances between the frame and ultra-light six-ounce silk tubulars blown up to over 130 pounds per square inch. (Six-day bikes have to be made with even stronger frames and shod with cotton tyres.) Match sprinting takes place over a distance of 1,000 metres, and for such short distances the rider can afford to adopt an extreme position, with a high saddle and strong, deep bars set low on a downward-sloping stem. The track bike is the simplest and purest speed machine in the cycling stable, and specialist hill climbers and some time triallists favour them on the road as well. For pursuit races the rider will keep to the inside edge of the oval and his effort must last longer and be more evenly sustained. Thus a more comfortable position must be adopted and a lighter and less robust machine can be used. The East German pursuit team at the Montreal Olympics used lugless frames, smoothed off wherever possible, and specially flattened radial spokes.

Track events are frequently classified as 'entertainment' or 'championship' races according to whether they are part of the Olympic and World Championship calendar. Both categories demand much in skill and courage. Let us look first at the world of the six-day race – a unique and exciting blend of sport and entertainment that shows very clearly how much track racing has to offer and explains why it has been so popular in the past.

Consider the Skol Six-Day at the Empire Pool, Wembley, in London, the Ghent Six in Belgium or the Munich Six at the Olympic Stadium. During the evening a variety of races will be held, including sprints, pursuits, tandem races and perhaps some motor-paced work. The centrepiece is the Madison, a race for as many as twelve teams of two riders, which is staged on each one of the six days. This is the ultimate in professional spectacle as points, prizes and laps gained are accumulated from each race for almost a week – a competition in instalments like a TV serial.

These events have a long and colourful history going back at least to Birmingham in 1875, when riders set out to pedal Ordinaries round an outdoor track for twelve hours every day. American promoters took to the idea and by the 1890s 'Sixes' were established as endurance events in which the racers had to do lap after lap for twelve, then for eighteen, and even for twenty-four hours a day with only brief rests for sleep and massage. In 1897 C. W. Miller achieved

John Nicholson maintains rear position in a match sprint standstill. He will not be forced to take the lead.

the all-time record distance of 2088.35 miles in a New York Six. This is an astounding average of 14½ mph for every hour of those six days. He slept for only 4½ of the 9¾ hours he spent off his bike. Although these grim feats were popular with the large crowds they did not offer much variety. Thus professional women riders came to be featured and all-women races were also introduced to add novelty and glamour. The girls would ride for two to eight hours a day dressed for competition in divided skirts or knickerbockers and voluminous blouses.

The early American Sixes were still rather grim affairs, despite the ladies. There was constant noise – from the crowd, from the bands hired to play and from the megaphones used for public address, all crammed into a dusty and smoke-laden cavern. The managers would bully their men into continuing even after the inevitable collapses and collisions. The riders who finished ended by begging for sleep, dazed by their efforts, hallucinating or hypnotized by what the press liked to call 'the race to nowhere'. Public opinion began to turn against such spectacles and a law was passed restricting rides to twelve hours a day.

In 1899 two-man teams were finally introduced at Madison Square Gardens in New York so that riders could rest while their partners took a turn. The 'Madison' style caught on and spread to Europe where it is still known as *l'Américaine*. Six-day racing was extremely popular in the United States until public interest finally died away by the 1930s leaving the field to the fans in Europe. Over the years the brutal hours of endurance had lessened, and the points system introduced in the twenties gave variety and change of pace to the programme. Even so, until as late as the 1960s riders were expected to be on the track for many hours. You could visit a Six during a quiet period around four or five in the morning and see men casually circling the track with their handlebars turned up, wearing warm clothes and steering with a slippered foot as they read the newspapers or had a bite to eat.

The modern Six is a much more humane affair and within eight hours, often from mid-afternoon to midnight, it presents a varied and thrilling programme. The Madison is the race that links the lot together with a session or two every day. The aim is to gain laps and to accumulate points. The points are awarded for places in the sprint for the line which happens every ten laps throughout the race. If a team manages to gain a lap on the rest then this is enough to ensure victory, otherwise the placings are decided on points. Some teams will go for points alone, as there are rewards for this as well. As many as twelve riders can be racing on the track while their team-mates circle slowly above them on the banking. As a man tires or completes his agreed stint, his team-mate will drop down into the race and take over by getting a 'sling' from his opposite number. This is done by gripping the replacement's hand and flinging him forward into the fray, or by using a special pad sewn into the hip of his shorts.

Typical tactics depend on the strengths of the riders concerned and their place in the competition. A team might consist of a sprint specialist paired with an all-round pursuit rider. This is a common Madison ideal, realized, for example, in the successful partnership between Patrick Sercu and Eddy Merckx. The sprinter will be the one who rides for the points every ten laps. If a rival team plans to gain a lap, however, they might attack with their pursuiter immediately after their opponents have given everything to the sprint. But other riders may have noticed that the rival pursuit man 'sat in' for an easy ride during the sprint, and so they will be on the alert for

Masters at work: Peter Post hand-slings Patrick Sercu in a Six at Cologne.

his subsequent attack and quick to counter it or to join him in the breakaway. The tactical complications can be imagined, especially when teams begin to enter coalitions to try to frustrate the points leaders or to help each other regain a lost lap.

The excitement is at its most intense during an attack, when riders swoop up and down the bankings exchanging places every lap or half-lap in a glittering confusion of bright silk jerseys and drumming wheels, twenty-four men weaving and 'jamming' hard while the crowd cheers and boos and the commentator pours out a breathless summary of the action and the changing score. When the time or the distance is completed the race dies down until the next instalment and the programme moves on to other events, perhaps an amateur sprint match or a motor-paced solo effort against the clock.

Professional track racing in six-day and Omnium events is most common during the winter season in Europe, especially in West Germany, and many riders – road men too – earn their best living this way. Commercial sponsors and local firms will buy advertising space around the track and on its surface, and various special prizes contribute to the glamour and the rivalry. Awards of motorcars are common, and, indeed, some riders concentrate on bringing home the goods rather than on going for outright victory.

In some velodromes the spectators sit in the central area and eat

A Six begins at the Sport Palace in Madrid.

and drink as they watch the action flashing around their tables. In keeping with this atmosphere, certain professionals become well known to the crowd for their clowning and cheeky skill. Tom Simpson, the brilliant British pro for Peugeot, used to appear before Continental audiences sporting a bowler hat and an umbrella – 'Major Tom', they called him. Peter Post, the big Dutchman who now manages T I–Raleigh, dominated six-day racing in his time, and he would create thrilling motor-paced 'finishes' by hanging back until it was almost too late. There is plenty of scope for entertainment. The 'Devil', for example, is an elimination race that whittles the field down by removing the last man across the line on each lap until only a few are left to fight it out at the end ('devil take the hindmost'). Some riders will play at nearly being caught out each time, only to sprint past the 'hindmost' at the last minute, blowing a whistle or honking a horn and waving him off grandly. The crowd loves it.

A good Six is a splendid occasion and they deserve to be better known than they are in Britain, where the fans have to be content with the very popular Skol Six in September each year. One of the problems is that velodromes are expensive and there are too few of them to train young riders and to give spectators a chance to see the sport in the best surroundings. Track events are very popular in

Australia, where they are characteristically tough and fast. Indeed, some of the best sprinters in recent European competition have been Australians. Cycling on the boards thrives particularly well in Japan, too, although the Kierin meetings have less variety than their Western counterparts. Japanese competition concentrates on straightforward races to the line with a field of nine or ten riders. Commercial sponsorship and a system of tote betting makes it a very well-attended and wealthy sport. The Japanese ignore the wary punter's advice – 'Never bet on anything that can talk' – for there are 3,000 track professionals who make a good living from their wheels. International talent was bound to emerge soon from such fertile ground. In the 1977 World Championships, pro sprinters Nakano and Sugata won the gold and silver medals.

In the United States for the first thirty years of the century, track sport was very popular and American riders, sprinters especially, were among the best in the world. As well as Zimmerman, who rode for Raleigh, there was Frank Kramer who was his country's professional sprint champion for eighteen years. Kramer was a dedicated athlete whose spartan life style was much more in the modern idiom than Zimmy's taste for late nights and cigars. Then there was the famous black rider Marshall 'Major' Taylor, who overcame prejudice and the very toughest competition at home and in France. Taylor's career was remarkable. He began by sweeping up in a bicycle shop in Indianapolis and ended by writing an autobiography entitled, with justification, *The Fastest Bicycle Rider in the World.*

In recent years there has been a revival in U S track competition, with more than a dozen velodromes available around the country. American women riders in particular have done well on the international scene, winning six World medals in seven years in the face of what was once virtually a Russian monopoly. They have done a lot for American cycling, and for women's racing in particular. Surprisingly, there is still no category for women riders in the Olympics. This should be changed.

This brings us to the 'championship' events, which are closer to the world of pure track athletics. These comprise the kilometre, the pursuit, the sprint and, finally, the motor-paced races which are close in spirit to the epics of the old six-day scene.

The *kilometre time trial* is a race against the clock in which each man takes to the track alone for little more than one minute of all-out

effort. From a stationary start the rider has to accelerate to his full pace and maintain it to the end without fading. In the '76 Olympics Klaus-Jurgen Grunke of East Germany won the Gold medal with a magnificent 1 minute 5.297 seconds, an average of 34.237 mph from standstill. Only one tenth of a second separated the Silver from the Bronze medallist at Montreal.

The *individual pursuit* is also a test of cycling power and judgement, as the rider has to plan his effort so that he finishes at full tilt without an ounce of energy left. Yet this event includes all the psychological pressures of racing against an opponent. The two riders start at the same time but at opposite sides of the track. A light flashes as each man passes his starting line to let the spectators see which of them is ahead. The distance is four kilometres for amateurs, five for professionals and three for women, and it is not unusual to see the lead change several times as riders find their full stride or go through bad patches. The race stops early if one man catches the other, but usually it is decided by a matter of only fractions of a second. A totally economical yet powerful and streamlined style is vital in this event: nothing should move on a good pursuiter but his legs as he whirls relentlessly round, trying to gain an inch and then two inches on his rival. British pro Hugh Porter has excelled at this, winning four world titles and a nickname – 'the Locomotive'. It is a race that takes a strong mind as well as muscle.

Team pursuiting is on the same lines, except that it demands four riders who are very well matched and disciplined. Each man takes a turn at the front, usually for only half a lap, and then soars up the banking as he reaches the end of the oval to drop down into his place at the back of the line in the shelter of his team-mates. The group's time is judged by the third man, so it is not unusual to see the strongest rider 'blow himself up' on the second last lap by taking a full circuit flat out before dropping back, exhausted and out of the race. A well-trained pursuit team is an impressive sight, riding in perfect line and matched rhythm with only inches between their wheels as they slot in and out of place.

The *sprint* is the most subtle form of track competition because it is here that riding in your opponent's slipstream offers most tactical advantage. Sprint racing is run as a series of eliminations, sometimes starting with three men at a time but always moving towards the classic two-man confrontation, side by side at the starting line. The best riders eliminated in the earlier rounds can make their way back into the running through a series of *repêchages* – 're-sits' or 'scrape-throughs'. The sprint distance is a thousand metres, which is only three laps of an official championship track, but it is not a simple test of speed. A rider will try to tuck in behind his opponent and follow him to the home straight. Having conserved his energy in this way he flicks out at the last second and drives his machine past the leader.

Sprint finals are run on a 'best of three' basis, with riders drawing lots and then taking turns to start on the inside of the track. This is because the inside rider is obliged to take the lead for at least the first

Precision pursuit racing with four riders all on the same pedal stroke.

lap, unless his rival elects to do so. Spectators can expect to see a game of cat and mouse as each rider manoeuvres for position or psychological advantage. Sprinters will bring their bikes to a complete standstill in the attempt to force the other man to take the lead, or they will climb and dive on the banking to test his skill or do a standstill on the steepest part to test his nerve. Speed is not important at this stage, as only the last two hundred metres of the race are timed. But when a sprinter does make a break for it he puts all his considerable strength into the 'jump' – a burst of high-speed pedalling that takes him towards the line at forty miles an hour, head down, back arched, in an explosion of total effort. Sprinting is a tough sport, especially among seasoned professionals, and it demands courage as well as great bodily strength as riders rise and dive on the slopes or charge shoulder to shoulder, elbow to elbow, inches apart for the line.

Leading sprinters somehow have the ability to put in a second jump of acceleration, even when apparently riding flat out. Britain's

Reg Harris comes from behind to beat Dutchman Arie Van Vliet. The rivalry between these two thrilled huge crowds at home and abroad throughout the early fifties.

Reg Harris won many races in this way. He was an enormously strong rider, and his iron discipline and temperament are legendary. In 1948, for example, he had a bad car crash which doctors believed would put an end to his cycling career for ever. But he overcame a broken back and partial paralysis of the legs to win, in the following year, the first of his four professional world titles. At the age of fifty-

A tandem standstill is a rare test of balance.

one Harris came back to pro racing as the result of a challenge, and he took the Bronze Medal in the 1971 British Championships. He created an even greater stir when he returned again to take the Gold in 1974 and next year the Silver – an extraordinary feat for a man who had retired from cycle sport eighteen years earlier.

Motor-paced racing is often put on in six-day meetings, but usually small motorized bicycles called Dernys are used to attain speeds of about forty miles an hour. Championship motor-pacing, however, is the fastest, strangest and most specialized form of track sport. The 'stayers' ride behind large, old-fashioned-looking motor-cycles with engines of 1,000 cc or more and belt drive to give oil-free and smooth acceleration. Rollers are fixed to the rear to stop the cyclist getting too close for safety and to regulate the suction created by what is, in effect, a moving windbreak. The stayer wears a stout crash helmet and his bicycle is specially designed and strengthened to cope with the G forces of the speeds on the banking. His front wheel is smaller than usual and the forks are reversed to allow him to get as close as possible to the motorbike. The rider places responsibility for the tactics of the race firmly in the hands of his 'pacer', and

'. . . a moving windbreak'.

a clever and experienced man at the front can make all the difference. Indeed, there have been suspicions in the past that pacers have been persuaded to influence races without reference to the rider at all. Problems also arise if regulations insist that stayer and pacer be of the same nationality. Thus a rider from a small country may be deprived of an experienced pacer. On the other hand, there may also be room for doubts if the man on the motorcycle is the same nationality as his stayer's chief rivals. It is a complicated issue. It is strange to think of this as an Olympic event and even stranger to realize that in the World Championships the pacer gets a medal too.

Professional World Championship motor-pacing lasts for an hour at speeds of over forty-five miles an hour and the amateur event is over a distance of fifty kilometres. It demands a special kind of experience and nerve to ride at this velocity tucked behind a heavy motorcycle as it forges its way through the buffeting air currents created by the other pacers – as many as eight at a time roaring round the track with open exhausts.

In *A Moveable Feast* Ernest Hemingway recalls Paris in the twenties and the special atmosphere of the cycle tracks in the great days of motor-pacing:

. . . I have started many stories about bicycle racing but have never written one that is as good as the races are both on the indoor and outdoor tracks and on the roads. But I will get the Vélodrome d'Hiver with the smoky light of the afternoon and the high-banked wooden track and the whirring sound the tyres made on the wood as the riders passed, the effort and the tactics as the riders climbed and plunged, each one a part of his machine; I will get the magic of the *demi-fond*, the noise of the motors with their rollers set out behind them that the *entraîneurs* rode, wearing their heavy crash helmets and leaning backwards in their ponderous leather suits to shelter the riders who followed them from the air resistance, the riders in their lighter crash helmets bent low over their handlebars, their legs turning the huge gear sprockets and the small front wheels touching the roller behind the machine that gave them shelter to ride in, and the duels that were more exciting than anything, the put-put-ing of the motor-cycles and the riders elbow to elbow and wheel to wheel up and down and round at deadly speed until one man could not hold the pace and broke away and the solid wall of air that he had been sheltered against hit him.

. . . There were the programmes of the team races of two hours, with a series of pure sprints in their heats to fill the afternoon, the lonely absolute speed events of one man racing an hour against the clock, the terribly dangerous and beautiful races of one hundred kilometres on the big banked wooden five-hundred-metre bowl of the Stade Buffalo, the outdoor stadium at Montrouge where they raced behind big motor-cycles, Linart, the great Belgian champion that they called The Sioux for his profile, dropping his head to suck up cherry brandy from a rubber tube that connected with a hot-water bottle under his racing shirt when he needed it towards the end as he increased his savage speed, and the championships of France behind big motors of the six-hundred-and-sixty-metre cement track of the Parc du Prince near Auteuil, the wickedest track of all where we saw that great rider Ganay fall and heard his skull crumple under the crash helmet as you crack a hard-boiled egg against a stone to peel it on a picnic. I must write the strange world of the six-day races and the marvels of the road-racing in the mountains. French is the only language it has ever been written in properly and the terms are all French and that is what makes it hard to write.

It is a pity that Hemingway never wrote that book.

THE ROAD RACES

Road racing is for many the ultimate in excitement – the 'King of Sports' – and the lightweight road-racing bicycle is its finest all-round competition tool. It may be close-coupled for criterium races or rather less severe for long distances, but in either case it needs to be sturdier than a specialized TT machine and easy to control on any surface. On stage races the stars will have their gear ratios constantly altered to fit the country on the next day's racing, and they will change to specially lightened bikes for time-trial or mountain sections. The emphasis is on handling and rigidity, usually with

The road-racing bike in action with British pro Tommy Simpson. Simpson died tragically on the slopes of Mont Ventoux during the 1967 Tour de France. He was overcome by the heat and by exhaustion heightened by an unwise (and uncharacteristic) use of stimulants.

ten or twelve gears, fittings for a feeding bottle and a comfortable saddle. This has to be a machine for all conditions and many miles.

Both amateur and professional riders have a wide choice of events. There are single-day races from eighty to a hundred miles long, criteriums (town-based circuit races through the streets), and stage races demanding two or three weeks of effort across varied and changing terrain. Among the most famous amateur events are stage races like the British Milk Race and the East European Peace Race which draw leading teams from many countries. Ultimately, however, the finest competition comes from the thriving professional class in Europe, where the greatest riders and races have always been found. Recently there are signs of expansion in America and the UK, and the British pro season of 1977 looked particularly promising. After all, men like Seamus Elliott, Brian Robinson, Tom Simpson and Barry Hoban have already shown that British pros can have Continental class if they go abroad to compete wholeheartedly.

Professionalism and commercial interests were a driving force behind cycle road racing from its very beginnings. Paris–Brest–Paris was conceived to boost the sales of *Le Petit Journal* and the first ever race on the road, from Paris to Rouen in 1869, was sponsored by *Le Vélocipède illustré*. Six women rode in this event and it was won by James Moore, a famous amateur of his day. A new kind of rider was beginning to appear, however, often from a humble background and dedicated to success through hard training. There were early disputes over this growing professionalism and at the first Bordeaux–Paris race in 1891 Charles Terront and other leading French riders were not allowed to compete because the British amateurs objected. Nevertheless, despite a British victory the 'gentleman rider' soon faded from the European scene and professionalism was felt to be the obvious goal for a talented sportsman. The British continued to value amateur sport and moved on to develop time trialling. Even today amateur racing clubs in France and Belgium have much closer links with the pro scene than do their counterparts across the English Channel. Promising amateurs in these countries have supporters' clubs to help them finance themselves.

Team sponsors in the earlier years were cycle-makers like Alcyon, Peugeot and Mercier or tyre manufacturers like Dunlop and Wolber, but *extra-sportif* sponsors are now the rule, like Brooklyn who make chewing gum in Italy, Frisol Oils from Holland or Bic (pens). Many *extra-sportifs* join with cycle-makers or other firms to form alliances like Gan–Mercier (Gan is a life assurance company), Peugeot–Esso–Michelin or Ijsboerke–Colnago. Ijsboerke is Belgium's largest ice-cream maker. They spent well over £100,000 on their cycle team in 1976 and it rewarded them by winning the World Cup. The sponsor gains from such investments because his name and 'colours' are seen throughout Europe. Furthermore, the regional newspapers give massive free publicity in their sports pages every time a race is held in their district. In the absence of national dailies in the British sense this coverage is what particularly attracts the *extra-sportif* to cycling. The demand for publicity also encourages the star system under which the leading riders are given every support by their teams because big names make big news.

Expensive as it is, cycle sport is only a small part of the advertising budget for giants like Watneys (Europe) and BP. Indeed, the size of a few of these companies has led to bad feeling among some of the smaller sponsors more intimately connected with cycling, especially if the *extra-sportif* – like Fiat, say – can command vast amounts of money. Commercial alliances come and go, changing slightly from

'Publicity . . . and the star system'. Freddy Maertens crashes in the 1976 Paris–Roubaix. The cameramen are there before he is even released from the toeclips. Later in the year he will be World Champion.

year to year like the membership of their teams. Investment on a large scale, hundreds of races on the calendar, fields of up to 150 men in a race – all this explains a lot about the dominance of Continental cycling.

The professional year on the Continent takes in no less than sixteen 'Classics' – established one-day races of over 150 miles, eleven of which are 'World Cup' events which count towards an award made to the highest-placed trade team. There are at least 140 other important races between February and October, not to mention many smaller events. The oldest and most famous Classics have each developed a special character for the cycling enthusiast. Let us glance at a few highlights from a typical year. Milan–San Remo (founded in 1907) begins the classic season in March with the 'Race to the Sun' over almost 180 miles to the coast of the Italian Riviera. At the very end of the route the road rises to overlook San Remo before sweeping down to the finish. This climb, the famous Poggio, has been a killing ground for many years and Eddy Merckx attacked here in 1976 to win the race for a record seventh time.

Even in April the Paris–Roubaix race (established in 1896) seldom gets good weather. Riders fight their machines towards the velodrome at Roubaix, shattered by their battle through narrow and muddy cobbled lanes between the exposed beet fields of Flanders and Northern France. Roger De Vlaeminck set a heroic pace in 1976 but lost to Marc Demeyer and Francesco Moser in the last few metres. Only 38 out of 154 men finished, but the 'Hell of the North' is proud of its status as a test of mechanical and physical survival. It is from such a hard background that Belgium produces so many strong *coureurs.* De Vlaeminck managed to win in convincing style during a hot, dusty day the following year and this gave him a record four victories in the most gruelling one-day race of them all.

The longest Classic is the Bordeaux–Paris, a unique event held in May. The riders leave Bordeaux velodrome in the early hours of the morning to ride north for 358 miles right across France into the dawn. Around the half-way point they are met by small motorcycles and paced to the finish at the French capital. Relatively few riders enter this 'Derby of the Road' and those that do can expect to spend some fourteen hours in the saddle. Walter Godefroot, a specialist behind motors, won in 1976 with an average speed of 26.6 mph.

De Vlaeminck – 'the Gypsy' to his fans – attacks to take his record victory. 'Hell' was uncharacteristically hot and dusty in 1977.

The great national stage races dominate the calendar in the summer months: the Vuelta a España, the Giro d'Italia, and then the oldest and most famous of them all, the Tour de France. The World Championships come in September with separate races on a road circuit for both amateur and professional classes. The commercial teams break up for this fixture and the pros ride to represent their various countries. The winner must overcome considerable opposition, but he is not necessarily the best road racer of the year. For this distinction the Super Prestige Pernod Trophy is awarded according to points gained in a selection of the top-rated events. Young Freddy Maertens finished a splendid road season in 1976 by taking both titles. Finally, October brings the international time trials, and in 1976 Maertens led again in both the GP des Nations and the Baracchi. The closing one-day Classic of the year is the Tour of Lombardy, which brings the riders and the reporters back to Italy where it all began in March. Roger De Vlaeminck won this 'Race of the falling leaves' in 1976.

The pros have a long and full calendar and the continual stress of competing in what many say is the most physically demanding sport in the world sometimes leads to drug abuses. This is largely prevented by an elaborate system which tests all winners and takes random samples from the field, but cases of 'doping' still turn up from time to time. The issue is not always a simple one, since even the mildest over-the-counter cold cure can contain substances which will show 'positive'. Some unfortunate riders have been fined for medicines taken under a doctor's supervision. Deliberate doping is physically dangerous, nonetheless, and it probably serves as a psychological prop rather than anything else, for there is doubt as to whether it directly increases performance. But the psychological pressures of pro racing should not be underestimated.

The colour, complexity and history of Continental cycle racing reach their ultimate peak in *La Grande Boucle*, the Big Loop – the Tour de France. It was founded in 1903 by Henri Desgrange, the chief editor of *L'Auto*, who seized on the idea for publicity. Desgrange was no mean cyclist himself, an early hour record-holder and an ex-professional, and his tough-minded rulings on the Tour were famous, or notorious, during the thirty-two years he was *Directeur de la Course*. The 'father' of the Tour also conceived the publicity caravan which would accompany the racing men during the big loop along with cars for the race director and others. The whole affair was to move through the remotest parts of the countryside like some spectacular travelling circus. It was a greater success than even he could have foreseen, although sabotage from over-enthusiastic fans nearly brought the series to a premature end. In ten years the sales of *L'Auto* grew from 14½ million to an astonishing 43½ million in 1914, ample witness to the selling power of sport. This stresses once again how vital the connection between the press, advertising and the manufacturers was to the growth of cycle racing.

The heart of the Tour lay with the physical feats of the riders,

At the end of the race for the rainbow jersey, these faces say everything about the efforts involved in professional road racing. De Vlaeminck, Zoetemelk and Merckx after the 1975 World Championship. They finished second, fifth and eighth.

however, and it was these that truly fired the public imagination. By 1906 the race had expanded to thirteen stages over nearly 3,000 miles. The riders were on the road for fifteen hours or more and exhausted men would still be finishing a stage hours after the leaders had crossed the line. No wonder they needed alternate rest days

Speicher (right) leads at an Alpine ascent during the 1936 Tour. Primitive enough then, in the really early years these roads were scarcely passable dirt tracks.

and no wonder more than two thirds of the field used to drop out before the end! Surprisingly enough, up until the twenties many riders were *touriste-routiers* who did not belong to teams and had to arrange their own accommodation in the stage towns at the end of each day.

The Tours were epics of endurance from the start. Riders had to change their own tyres if they punctured: the quick wheel-changes of today were not permitted until the 1950s. It was 1923 before Desgrange allowed cyclists to change damaged parts on their machines, and even so they were limited to one bike and had to make the repairs on their own. This helps to explain why they favoured the two cogs on their back axles rather than trust to the frail derailleurs of the time. Photographs of these heroes show them with heavy jerseys, bulging pockets and spare tubulars crossed around their shoulders. With blackened faces and sweat-streaked goggles, they look like survivors from some lost desert army, fighters in regions the rest of us know little about.

The modern Tours are faster – on better roads and with much tighter team tactics – with twenty or more stages over at least 2,500 miles. But they remain true to the spirit of the early days. The race is now organized and sponsored through the newspapers *Le Parisien Libéré* and *L'Équipe* (a daily entirely devoted to sport). M. Jacques Goddet is *Directeur de la Course*, the direct inheritor of the Desgrange tradition. The route changes slightly each year but it always describes a loop around France to finish in Paris. To convey something – only a little is possible – of how modern Tours are fought, let us follow the highlights of one particular race.

THE TOUR DE FRANCE 1975

This, the sixty-second edition, was one of the hardest and hottest Tours for many years. The prize money totalled £80,000 and the budget for the whole promotion was more than ten times this sum. The winner shares all his prize money with his team riders, for he could not win without them. He can look forward to the flood of contracts and appearances which will follow his victory. The other main awards are Points Leader and King of the Mountains, but there are also many secondary bonuses. To keep the action going, various prizes and points are awarded to the first riders across a line drawn on the road along the way. *Primes* are fought for in the high mountain passes. 'Hot spot' sprint prizes are sponsored near towns and villages. There are daily awards for such qualities as 'elegance' and 'combativity' and 'bad luck'. Thus the largesse is widely distributed – and, after all, it is no mean achievement simply to finish a Tour. There were only two rest days in twenty-three days of racing in the 1975 promotion and Bernard Thevenet, the overall victor, won by amassing a lead of 2 minutes 48 seconds after 114½ hours in the saddle. By the end of the race fifty-four of the 140 starters had dropped out and the last man home, affectionately known as 'red lamp', *la lanterne rouge*, had in the course of three weeks gradually lost just over three and a half hours to the leader in the yellow jersey. These figures give a good idea of how closely the modern Tours are contested and this particular race was hard fought indeed. It was a 'race of balance' between open-road riding, four time trials and four mountain-top finishes.

The *départ* is at Charleroi in Belgium, where the Tour begins with the Prologue, a time trial over 3.8 miles. Last year's winner, Eddy Merckx, wears the yellow jersey and thousands of his country-

Les vainquers du Tour: the first five finishers in the '75 edition. In the front row, from the left: Felice Gimondi, Bernard Thevenet, Lucien Van Impe, Joop Zoetemelk, Eddy Merckx.

men and women have come to Charleroi to see him keep it. But the short and twisting course around the suburbs and over tram lines produces an immediate upset, for the favourite can only manage second to a superb ride by 24-year-old Francesco Moser. A king has been deposed in little more than eight and a half minutes' racing. Moser is only in his third season as a pro, but already he is Italian National Champion and a rider to be reckoned with. There is a slight delay while officials remove the team name 'Molteni' from the first yellow jersey of the '75 Tour: they had been very sure that Merckx was going to win the Prologue. At last Moser exchanges his red, white and green Champion's colours for the *maillot jaune*, the badge of overall race leader. It bears the initials HD on the left breast in memory of Henri Desgrange and it is yellow because *L'Auto* was printed on yellow paper; it has been a feature of the Tours since 1919, much copied by other stage races. Moser has gained one of the most coveted distinctions in cycle sport and yet it carries a burden with it, for the rest of the field will now keep a close watch on his every move. His Filotex team will do all it can to help him keep it.

A bystander with a hosepipe provides welcome relief for the bunch.

After the sensations of the Prologue the first stage gets under way. This year it is in two parts, ending up 125 miles away at Roubaix, just inside the French border. Van Linden, in the 'white sky-blue' colours of Bianchi, wins the finishing sprint at the velodrome; but since Moser does not lose any time overall he retains his lead and collects a second yellow jersey for the next morning.

During the following four days Moser still wears yellow as the Tour forges across the baking roads towards the small resort town of Merlin Plage. Riding is competitive because the leaders are intent on gaining time if possible before the mountains come. But no one manages to break away on this fairly flat section of the route, for the bunch moves fast as riders take turns to make the pace at the front. The rest are arranged across the road in echelons for shelter behind the leaders. A road-block of mobile police keeps traffic clear. The heat rises and shimmers above the tarmac and bottles of mineral water are passed around among the cyclists. The men eat on the move too. As they sweep through predetermined feeding stations each rider snatches a little cotton satchel from the helpers and slings it round his neck. He will eat and drink from the *musette* as he rides or distribute its contents around his jersey pockets for later. Attacks

Like the heat, the camera is everywhere.

are sometimes planned at these stations so that a group can break away while the others are distracted. But everyone must eat sometime, for without regular food – fruit, rice-cakes, honey sandwiches and so on – the rider will fail for lack of 'fuel'. The British call this attack of nauseous weakness 'the Bonk' and all cyclists, from racers to tourists, learn to avoid it.

Life in the bunch is swift and crowded, requiring considerable skill in bike-handling. It is rare for machines to touch wheels, but sometimes a crash does happen and men go down like skittles. They

A pile-up at the rear. The rest of the field has disappeared up the road.

have to ride very hard indeed to regain the shelter of the *peloton*. Attacks can happen at such moments too, especially if one of the riders brought off is among the leaders.

Typically this kind of racing ends in a mass sprint in which all the front riders are credited with the same time. Rik Van Linden is a specialist with a very strong finishing burst and he is never lower than fourth during this period of the race. It follows that he too is closely marked, and in the fourth stage he falls at full speed in a clash, shoulder to shoulder at the line. Still, he takes second place and comes home second again, despite his injuries, in four out of the next five stages. Thus he gains the green jersey of Points Leader and establishes such a firm hold on it that he will keep it for the rest of the race and win the £1,000 Points Prize at Paris. Unfortunately for Van Linden no time bonuses are being given to stage winners this year. The *maillot jaune* is well advised not to contest such sprints too

Van Linden lies still at the end of Stage Four as the bunch sweeps past. Tangled in his machine, he looks as if he has fallen from the sky.

fiercely – an accident to him in the hurly burly could turn out to be just too costly.

Merlin Plage, like the other towns, has competed to be chosen as a *ville étape* – some have paid as much as £10,000 towards the Tour's expenses for the privilege. When the Tour arrives it brings huge crowds to watch and a convoy of over five hundred cars for journalists, officials, team mechanics, managers, doctors and masseurs. This means that over a thousand people travel with the Tour – an entire village on wheels. Organization on such a scale presents enormous problems and hotels will have been booked up by the public months ago, when it became known that the race was coming here. Merlin exists only as a holiday town and it is getting its money's worth this year, for Stage Six – a flat 12½-mile time trial – is held here as well, and 100,000 spectators have turned out to watch it.

While Moser has retained the yellow jersey and Van Linden's

sprint finishes have ensured him plenty of points, Eddy Merckx has been only two seconds behind the young Italian and he has not allowed Francesco to increase his overall lead. It is on this, the second time trial, that Merckx, wearing the rainbow jersey of World Champion, will get his revenge. In a ride that takes only 19½ minutes at over 30 mph he beats a fourth-placed Moser by 33.1 seconds and takes the overall lead by 31 seconds. He dons the yellow jersey and will keep it for the next eight stages.

After a day in the sun the thousands of spectators flock into Merlin Plage to sample the atmosphere of a stage town. In the evening a cinema screen is set up to show highlights of the race and rush shots of the day's action. For over 250 journalists the day's work is just beginning as they start to write or phone in their reports of Merckx's attack by the seaside. There is a lot to say about this strongly competitive and taciturn phenomenon whose presence has dominated Continental road races since 1969. The press is quick to make stars and quick to criticize them as well. The public likes to read of quarrels among the mighty and of a winner's failings. (As rival *campionissimi* in the late forties Bartali and Coppi endured much in the way of publicized feuds, and even Fausto Coppi's private life became the subject of scandal in the Italian newspapers.) Merckx won fewer races than usual in 1974 and his health has been called into question. Lately he has also had to cope with anonymous threats to his family, for he is a very wealthy and famous man. Yet he has never been afraid of pressure and if he wins this Tour, as he is surely determined to, he will have six victories to his name. This has never been done before – Jacques Anquetil was the nearest with five. Francesco Moser is also interviewed, now wearing the white jersey awarded to the leading young pro of the race. What are his plans? Will he counter-attack? There is a lot to write about.

Five days later, after a rest day, Stage Ten finds the Tour beginning to climb into the Pyrenees at the Spanish border. The race enters a new phase as the road rises and the riders begin to breathe hard in the ninety-degree heat. There are six major passes and three mountain-top finishes facing them before they arrive at the summit of the Puy-de-Dôme near Clermont-Ferrand. Then there is the second rest day.

Now the climbers come into their own, men like little Lucien Van Impe, who arrives first at enough summits to claim the King of the

Moser wins the sprint finish on Stage Seven.

Mountains classification. The white jersey with large red polka dots will remain his until the end of the race, when he will also collect a prize of £1,000. The deceptively frail-looking Dutchman Joop Zoetemelk begins to come forward on the slopes, winning the eleventh stage and finishing in the first four for the next six stages. He gains enough time here to secure his fourth place overall in Paris. And 27-year-old Bernard Thevenet also has plans for the mountains. This tough Frenchman is not a natural *grimpeur* but he believes that Merckx no longer has quite his old strength in the climbs and he and the new *directeur sportif* of a loyal Peugeot team are determined to prove it. At the moment he is in third place overall. Moser is starting to lose time in the mountains and, although he usually regains it by being a swift descender, a bad crash back at Pau has lowered his form for the moment. If Francesco is beginning to relinquish his claim on the lead, then sprinter Van Linden is soon struggling simply to stay in the race. His Italian team-mates are working hard for Bianchi as they cajole and even sometimes furtively push the green jersey towards the summits.

The *domestiques* of a team ride to protect their star. They will fall back in relays to help him catch up when he punctures, or exchange bikes if necessary; they fetch food and drink for him or lead out on suicide attacks with him safely in tow. These selfless 'water carriers' are indispensable on the modern Tours and many of them are very fine riders. It happens sometimes that even the lesser lights can have their day. Stage Twelve, for example, is a long, relatively flat ride in between mountain stages and the top men are content to leave it at that and to save their strength. It is Gerrie Knetemann's mother's birthday today and the Dutchman has a word with the various leaders before he makes a break from the bunch. They are content to let him and a few others go because they cannot threaten the overall positions. Knetemann wins the stage and Van Linden regains form by coming fourth in the sprint at the end. In the same way riders are sometimes 'set free' to go through their local towns in the lead so that they can stop and have a word with their families before the race catches up. But they had better stop. It would, in any case, be very difficult for a lone rider to hold off an infuriated *peloton* for long.

For the last effort before the rest day the riders face a finish at the top of the Puy-de-Dôme. The mountain is, indeed, dome-shaped – an old volcano that offers a fast, hard climb with over 3,000 feet to ascend in the last eight miles of road. Huge crowds line the route to see the spectacle. They have been picnicking there since the early morning, entertained by the noisy publicity caravan of vans and floats advertising and giving away their wares. Among other oddities, a giant bottle of Perrier water has driven by like some mirage at seven thousand feet up the sunny mountainside. The road is painted with the names of favourite riders and exhortations to do well. '*Poulidor*' and '*Allez Poupou!*' are much in evidence. Pretty girls have paraded past on motorcycles and loudspeakers are now reporting on the progress of the approaching riders.

The team cars and the organizers are kept in constant touch with

De Schoenmaecker leads for Merckx at 5,000 feet in the awful heat of the Pyrenees. The yellow jersey is beginning to look vulnerable and Zoetemelk will soon come from behind to cross the line first at yet another mountain-top finish. De Schoenmaecker worked like a hero for his leader on this and the following days.

the changing patterns of the race through the mobile transmitters and receivers of 'Radio Tour'. The journalists also pick this up in their cars as they follow the action from in front of or behind the main bunch. The Tour is regularly reported on national radio networks and this year Jacques Anquetil is doing the commentaries with a keen appreciation of the tactical subtleties. Helicopters chop through the air above the winding road and intrepid TV men use hand-held cameras from the back of motorcycles. Millions of viewers will see in close-up how each rider copes in his private battle against gravity and pain. Like gladiators on wheels the *coureurs* occupy their own quiet space at the centre of a moving ring of motor vehicles and faces. It is like this everywhere the Tour passes.

The excitement increases when the road is cleared by police outriders and the people strain their necks to catch a glimpse of the cyclists as they struggle upwards towards them. Van Impe passes and then Thevenet; they have saved their attack for the last hard miles. Merckx follows, hanging on grimly to defend the yellow jersey. Suddenly a spectator lurches forward and seems to punch Merckx hard. The Belgian is doubled up and badly winded, but somehow he manages to stay on his bike to reach the finish line in third place. It is a sensation. People mill about asking each other what really happened and get a dozen different answers. The incident is not explained; it seemed to be a wilful assault but the spectator claims that he was pushed. Cynics might note that he was a Frenchman and that feelings run high about what is, after all, clearly the greatest cycle race in the world. (Months later the gentleman is fined £75 at Clermont-Ferrand and Merckx asks for a symbolic one franc damages.) But for the moment the stage is over and the riders will soon be flown to Nice, where it will all start up again after only a day's rest. They have been racing now for two weeks and there are eight more days still to come.

In the last two stages eleven men have abandoned the struggle, including Jolliceramica's leading rider Knut Knudsen, a Norwegian who is a star pursuiter in his own right. The whole team will subsequently cut its losses and withdraw. It is sad to see Spanish climbing star Luis Ocana also drop out with knee trouble. Merckx has kept the yellow jersey for Molteni, but his lead has been diminished and the worst climbs and the Alps lie ahead.

During the rest day at Nice the riders are silent and withdrawn – each man thinking of the ordeal to come and knowing that he will have to dig deep into his reserves of courage and strength. Tours have been won and lost in the mountains and everyone suffers there. This knowledge and the effort involved can be seen in their faces. In the mountains a man will burn over three times a normal daily requirement of calories, and he may end up twelve or sixteen pounds lighter. On the climbs he will spend hours in 'oxygen debt' and his heartbeat may rise to 180 or 200 a minute – enough to cause a stroke in a less fit man. It is not uncommon for riders to need an oxygen mask after a hard finish at altitude. The *coureur* will move from high summer on the scorching lower roads to cold, mist-soaked and icy passes. At the summit he will pack plastic sheeting under his jersey

Merckx at the finish after the punching incident.

and pull on a pair of gloves before dropping into a freezing descent at speeds of over sixty miles an hour, inches from the edge of a narrow road as it winds down to the valleys below. The history of the Tour tells of terrible crashes and feats of courage on the descent, but fatal accidents are rare. In such conditions the most skilful bike-handlers will often catch up with the climbing specialists, at least until the road begins to rise again towards the second or third or fourth Alpine pass of the stage. No wonder their faces look pared to the bone at the finish line and their eyes seem distant, focused still on the summits of willpower they have crossed somewhere deep within themselves.

It is on the fifteenth stage that Bernard Thevenet wins the Tour de France for 1975. This is a hard climb out of Nice for 135 miles into the mountains of the Haute-Provence over five *cols* and four major descents. The mountain passes are classified according to their difficulty, with the first category containing the biggest ascents – those over 2,000 metres high. The route from Nice rises and falls, gaining altitude all the time until the last fifty miles, in which it crosses two first-category climbs and ends on a second-category mountain-top finish at the ski station of Pra-Loup. The leaders keep a close watch on each other throughout a hot day until they approach the last two giants, the Col des Champs and the Col d'Allos.

On the first peak Thevenet attacks twice, determined to overtake the yellow jersey with the help of his Peugeot team-mate Raymond Delisle. But Merckx fights back each time, supported by two faithful Molteni riders, Edouard Janssens and Jos De Schoenmaecker. Also in the leading group are Felice Gimondi for Bianchi, Luis Balague for Super Ser, Moser, supported by Joseph Fuchs for Filotex, Zoetemelk for Gan–Mercier and finally Van Impe for Gitane – seeming, as always, to turn his pedals swiftly and easily, hardly ever getting out of the saddle. The descent of the Col des Champs is typically rough with a poor surface and concrete drain channels running with melting snow across the road. Moser and Thevenet both puncture, but quickly exchange bicycles with their respective team-mates. They have to work desperately hard to regain the group just before the summit of the Col d'Allos. At moments like this even close rivals will help each other to get back into the running.

Thevenet is right: Merckx *can* be challenged on the mountains this year. Furthermore, the Belgian is in pain, for an old injury from a motor-pacing crash in 1969 is troubling his back. That punch could not have helped either. But the *maillot jaune* is fighting for his life and so he attacks again, still on the climb, and De Schoenmaecker exhausts himself to help his leader drive for the top and leave the group behind. The crowds at the summit roar with excitement. This will be the attack that wins the stage. Clearly Monsieur Eddy has lost none of his aggression.

Merckx is a particularly fast and daring descender and he is counting on 3,500 feet of precipitous Alpine hairpins to increase his lead before the final climb to the finish. He plummets down the road – scarcely nine feet wide – and suddenly comes upon a group of Press cars which has slowed down and dropped back to watch. At fifty miles an hour Merckx switches his bicycle between them, shouting furiously and shaking his fist. He comes through safely but he has lost a little time. Gimondi is a good downhill man and so he too swoops after the leader, leaving a tired Thevenet behind to descend with Van Impe and Zoetemelk – and to face the last slopes with two of the Tour's leading climbers.

These Alpine roads are frightening at speed and in fact the Bianchi team car crashes at this spot. Manager Ferretti and his mechanic only save themselves by tumbling forty feet to safety as their vehicle topples over the edge, scattering its shining load of bicycles and wheels into space. The officials at the finish hear confusing reports about what has happened. Have riders been killed? Was it the Bianchi car, or a *white* car? No one seems to know for sure and meanwhile the race is getting closer every minute. Every Tour has its incidents – always on the move.

The final stretch to Pra-Loup presents a crippling climb of 1,500 feet in only four and a half miles. But as this prospect looms ahead, Thevenet finds courage from somewhere and starts to attack once again, stamping on the pedals and forcing the bicycle upwards and away from his surprised rivals. Has he not been planning for the mountains? On the road ahead Merckx, who looked so strong, has suddenly found that he has nothing left to give and his legs feel turned to putty. It is a moment that riders dread. Gimondi has already joined him and left him labouring behind. Then Thevenet appears on the road below and the noise of the crowd blots out

With 3 kilometres and 800 feet still to climb, Bernard Thevenet recaptures an exhausted Merckx and starts to pass him. When he reaches the ski station at Pra-Loup he will have the yellow jersey. The faces tell all.

everything as they see the Frenchman force himself past the race leader. Zoetemelk and Van Impe are next to overtake the *maillot jaune* and commentators are to say that never has Merckx been seen to suffer so terribly. Thevenet gains strength now, manages to claw his way up to Gimondi and takes some relief by riding behind him for a spell. Then, with little more than a mile left to go, the Peugeot rider slips his bike into a bigger gear, moves alongside the Italian and forces himself into the assault again. Gimondi cannot quite 'catch his wheel' and he drifts to the rear.

The crowd senses what this means for their countryman; it screams encouragement all the way to the finish line and goes wild when he crosses it at last. Gimondi arrives 23 seconds later, then Zoetemelk 49 seconds after that, followed in half a minute by Van Impe; finally, 14 seconds behind him, a shattered Merckx drags in. Moser completes the leaders' board and he is over a minute behind the fifth man. The clock is ruthless: Merckx has lost 1 minute 56 seconds to Thevenet and the Frenchman has gained the yellow

Thevenet among the Izoard rocks stamps upwards to keep the yellow jersey for Bastille Day. The TV motorbike, the race director and the team car are right behind him.

jersey by 58 seconds. Bernard's face is delighted. The others have that distant look in their eyes. A little later the Tour doctor will find Gimondi 'positive' in the routine drug test and he will be fined accordingly. It will be a weary night for them all, for it is difficult to relax after such efforts, even in the face of exhaustion.

On the following day, 14 July, Bernard Thevenet wears the yellow jersey and defends his position brilliantly across the Col de Vars and the dreadful 7,500-foot Col d'Izoard. But not before Merckx has attacked – astonishingly – on the descent of the Vars. The Peugeot men ride as if possessed and finally bring their leader back to the breakaway. The crowds are already cheering for another French victory, shouting, '*Allez, allez! Nanard! Nanard!*' Out on the

desolate moonscape of broken rocks called the Casse Deserte, a young girl in a bathing costume holds up a placard which reads, 'Merckx, the Bastille has fallen, Thevenet has won!' It is Bastille Day. Not far away there is a cairn in memory of Fausto Coppi, the great Italian *campionissimo*, but it is Thevenet's day and he attacks so strongly on the Izoard that he leaves even the climbers well behind him. He crosses the line nearly two and a half minutes ahead of the rest of the field. The lead is secured and he will keep the *maillot jaune* for the rest of the race. The little stage town of Serre Chevalier sings and celebrates all through the night.

It was Thevenet's triumph as he rode towards the final circuit in front of half a million people in the Champs-Élysées, but it turned out to be Merckx's race after all. At the *départ* on the morning of 15 July the Belgian crashed in the 'neutralized zone', during the procession which brings the riders out of the crowded streets before the real racing starts. These accidents sometimes happen when wheels touch in the mêlée of a crowded bunch, and in fact Thevenet was brought down in a similar incident the next day. Merckx fell hard and got up dazed and vomiting but he ignored the doctor's advice to abandon. Later it was found that he had broken a cheek-bone and perforated a sinus. The doctor refused to take responsibility if he continued to race. The World Champion had his food mashed to help him swallow and carried on, looking like a boxer after a fight. He rodc the rest of the race in severe pain because analgesic drugs are banned. In the remaining few stages, with great courage, he even managed to reclaim 33 seconds from the yellow jersey's lead.

Merckx's refusal to give in graced Thevenet's victory and more than justified his own second place in the final general classification. This astonishing rider also came second to Van Linden in the Points competition and second to Van Impe in the King of the Mountains. Van Impe finished third overall, ahead of Zoetemelk, Gimondi, Lopez-Carril and, in seventh place, Francesco Moser, who had started it all off, three weeks and 2,400 miles ago. It was one of the good Tours.

7 The Bicycle in Fashion Again: Another Look

Every day, on the grounds of 'our own safety', we are effectively denied the freedom of the city. Pedestrian barriers, overhead walkways, parking meters, inner-city ring roads and lengthy one-way systems hem us in on every side. Surely this must be changed? Yet it is best not to become too shrill, for these are old and subtle problems. Some odd facts come to light, for example, if we look at the long-running legal and social wrangle of The Car versus Society.

Amazing though it may seem, more people were killed on the roads of Britain in 1934 than in 1963. In 1934, with only 1,300,000 cars, there were 7,343 deaths, much the same kind of figure as for the preceding five years; 1963 saw nearly six times as many cars (7,300,000) and yet 421 fewer folk were killed in road accidents. In April 1934 Parliament discussed the Road Traffic Bill, the first major legislative proposal to check the carnage. Lieutenant-Colonel

Moore-Brabazon spoke on behalf of car-owners, who saw themselves as a besieged minority about to lose essential and valid personal freedoms. He was eloquent in the cause:

> It is true that 7,000 people are killed in motor accidents, but it is not always going on like that. People are getting used to the new conditions. The fact that the road is practically the great railway of the country instead of the playground of the young has to be realized. No doubt many of the old Members of the House will recollect the numbers of chickens we killed in the old days. We used to come back with the radiator stuffed with feathers. It was the same with dogs. Dogs get out of the way of motor cars nowadays and you never kill one. There is education even in the lower animals. These things will right themselves.
>
> *Hansard* (10 April 1934)

But Parliament passed the Act just the same and 1935 saw the first reduction in accidents for several years – even if the British pedestrian had still not learned his lesson quite as thoroughly as the chickens and the dogs.

What were the freedoms which the motorist had lost? Speeds in urban areas (roads with street lighting) were to be restricted to 30 mph. Up until 1930 there had been a blanket speed limit of 20 mph, but it had been regarded as an obsolete technicality. Perhaps more revolutionary was the fact that the driving licence was now to be dependent on a test of driving skill, though established licence-holders were not to be examined. Previously it had been a simple matter of signing an unsupported declaration of medical fitness. In London hooting car horns was forbidden, first at night and then more generally. This was both to stop noise and to slow cars down, drivers being thus prevented from claiming that they gave 'fair warning' to pedestrians. 'Jay-walking' was to be forbidden, but the pedestrian crossing was 'invented'. These were to be signalled by orange flashing globes, much derided for the carnival appearance they gave to the streets and nicknamed 'Belisha Beacons' after the Minister of Transport who had seen the Act through Parliament. These innovations have all become basic fixtures of the British motoring scene.

Although the 30-mph restriction is scarcely observed today in many city streets, even by the police, few drivers would actively oppose it. Most countries in the world with fast roads and traffic problems used the fuel crises of the seventies to introduce general speed limits. These were first seen as economy measures but they were retained on grounds of safety, and there seemed to be general public indifference to this in spite of the enormously influential car lobby. Perhaps a gradual shift in public opinion is

The daily drive on the approach to Harbour Bridge in Sydney.

beginning, or maybe the roads are so clogged that no one expects to be able to use the modern car's high speed potential anyway. Clearly, however, road safety for all of us and the future of the motor vehicle in society ultimately depend on legislation and politics. Public opinion favoured the 1934 Act. Today, world oil prices influence speed limits. The dealings of countries thousands of miles from the motorways of Europe and America may yet forge new transport priorities which could eventually terminate the right to private motoring on which we have come to depend.

The bicycle faction has also involved itself in political lobbying, some of which now seems a trifle wrongheaded. In September 1939 motorcar headlights were forbidden as part of British wartime blackout regulations. Within a month deaths due to road accidents had increased by 100 per cent, and masked lights allowing minimal illumination were introduced. Not unnaturally in these conditions, rear lights for bicycles were deemed compulsory. Towards the end of the war the government decided to perpetuate this situation by making rear lights, reflectors and white patches mandatory equipment even after blackout conditions were raised. The CTC fought tooth and nail against a decision which it considered pernicious and even dangerous.

The main objection to rear lights was that the onus of avoiding accidents ought always to rest with the overtaker rather than the overtaken; motorists should adjust their speed to suit visibility and conditions and not depend on cyclists being self-illuminated. If cyclists had to carry lights, what of pedestrians? And one can see the

CTC's point. Every defensive move towards safety accepts rather than prevents the evil which it recognizes.

In these matters there must be a delicate balance between car and bicycle which needs careful consideration. The logically final solution to the defensive safety ethic would be to prevent all cycling and private motoring. (The only utterly safe bicycle is an exercise machine bolted to your drawing-room floor, and you should wear a crash helmet just in case you topple off that.) Certainly road deaths would be dramatically reduced. Such a measure would not be popular, of course, and so it is that transport planners, like First World War generals, must decide what is an acceptable number of deaths on the road. For 1973 in Britain this figure stood at 8,234, and in the USA

55,800 people died in the same year, almost 10,000 more Americans than had been killed in Vietnam in eight years of armed conflict.

In a world where the motor vehicle rules the roads, the bicycle lobby will have to fight hard to preserve any sense of balance. What is good for the car – faster roads, for example – is often fatal for the cyclist.

The safety equation differs from country to country. In the USA cyclists seem to have accepted the need to festoon their machines with reflectors, some of which would be illegal in Britain. The plastic crash helmet with air-scoops to cool the head has become accepted gear for American cyclists, while in Europe it is never seen on the roads and only occasionally on the track. Racers do wear a sponge-and-leather 'helmet' which may protect the head from scrapes but not from more serious damage.

There is a similar kind of contrast between American and British attitudes to 'bikeways'. The CTC approves of 'cyclepaths', which are usually country pathways far from roads and available only to the cyclist, horseman or walker. It also approves of properly organized systems of cycleways which avoid contact with the main roads altogether, such as exist in Stevenage. But it preserves its old antagonism to the kind of cycle track that runs alongside a car route and dumps the cyclist back into the main flow of traffic at the road intersections – just at the most dangerous points, in fact. Moreover, the CTC argues that the surface of cycleways is likely to be a low priority so far as maintenance is concerned, even though a bad surface is a greater danger to the cyclist than it is to the car-driver. Their fears have certainly been justified. Many cycle tracks were built flanking dual-carriageway roads all over Britain just before and after the last war. They are now so ill-repaired as to be unusable. Thus it is that the old guard of British cyclists will stick resolutely to the main road and disregard the cycle track with its new cosmetic blue and white sign.

Will the thousands of miles of American bikeways prove to have been a mistake? The craze of the 1890s saw some wild plans for the Safety, including the construction of bicycle paths on the updated elevated railway of New York, providing a whole network of aerial rides above the city streets, smoke and fumes notwithstanding. American cycling enthusiasts seem to have gone about proposing and constructing bikeways with equal fervour. Let us hope that they

Boston Common on Sunday morning.

have not created a situation which might make it easier for authority to ban the bicycle from the open roads altogether. The best plans should aim for fully integrated networks like the 155 miles of cycleway inside Amsterdam.

New York's Central Park became a symbolic focus for the cycling boom of the late sixties. During the fifties Dr Paul Dudley White, Eisenhower's medical adviser, had been at the forefront of a health-through-cycling campaign. But it took park riding to add radical chic. (Ironically, in Edinburgh police actually patrol the park pathways in cars to prevent cyclists from riding on them.) The American boom may have started as fun rather than as a real transport alternative, but it was still a significant event. Some sources estimate that by 1974 there were as many as 100 million bikes in America. Politicians latched on to the new pastime.

'Putting the bicycle back in the city,' said Secretary of the Interior Stewart Udall, 'is ten times more important than building the SST [super-sonic transport].' How typical of the bicycle to take on Concorde, as if the motorcar were not enough!

It remains to be seen whether cycling in the States will go beyond being a pastime. The boom, in retrospect, seems to have been a triumph for advertising and business but not necessarily a new vision of personal transport. Suddenly the bike, characteristically a middle-price-range lightweight 'ten-speed', became an essential possession. Bicycle-thieving also flourished on a vast scale. It would be interesting to know how many of those 100 million bicycles are now in regular use and how many are gently acquiring a patina of aluminium oxide in the darkest corner of a garage. The cycle industry found a market and sold hard, but the crest of that commercial wave has now passed. Nevertheless, the market in the United States remains large enough to attract confidence and innovation.

With its lack of cycling tradition, the American attitude to the bicycle is refreshingly radical. There has been a healthy desire to

The bike on the right is better suited to modern city riding, if only because it is less likely to be stolen. Its young owner has put together a neat runabout with alloy cotterless cranks and a three-speed hub.

experiment with the basic design of the machine and its components. Big companies are experimenting with improved designs for gear shifters and brake calipers; disc brakes, hydraulic levers and 'solid' tyres might soon be common. Elliptical chainwheels are on the market again, and a re-awakened interest in streamliners might lead to commercially available handlebar farings. Such variety and ingenuity is surely a good sign, and no doubt our old friend the anatomical saddle will be reappearing in due course. There also exists a melting pot of private inventors in the old style, as well as genuine eccentrics and what looks like a downright lunatic fringe. Twenty-five-speed bicycles, home-built modifications and recumbent cycles like the splendid machines of Captain Dan Henry are regularly featured in the American cycling press. At the same time, correspondents give alarming accounts of helmet-mounted revolving warning lights, electric socks and, for riding in very cold weather, a cut down gas-mask with the tube tucked into your trousers.

Radical perspectives on the bicycle go hand in hand with radical

politics, and the bike and moral rectitude to some idealists seem inextricably linked. Carl Bernstein discovered this problem, amongst others, while following the trail from Watergate:

> He had picked up a profoundly disturbing piece of information that day: Magruder was a bike freak. Bernstein had trouble swallowing the information that a bicycle nut could be a Watergate bugger. And Magruder really was a card-carrying bicycle freak who had ridden to the White House every day.
>
> Nobody could ever steal Jeb Magruder's bike, at least not there.

Bernstein and Woodward, *All the President's Men* (1974)

The morality of the bicycle is not just a pleasant assumption made by its enthusiastic riders. The bike has its theologians and political theorists as well. The most consistent and interesting of these is Ivan Illich. No simple autophobe, Illich advances a social theory that is revolutionary, and not particularly sympathetic to the problems of the over-developed Western world. The bicycle for him becomes

'Horsed upon the sightless couriers of the air' – Macbeth is quoting Shakespeare as he crushes the opposition beneath his wheels. Has he forgotten how the play ends?

part of an elaborate argument 'to show that two thirds of mankind still can avoid passing through the industrial age, by choosing right now a post-industrial balance in their mode of production which the hyper-industrial nations will be forced to adopt as an alternative to chaos'. For Illich the contemporary attitudes of rich countries to education, medical care and transport priorities are misguided, unprincipled and pernicious in their consequences:

> Transportation beyond bicycle speeds demands power inputs from the environment. Velocity translates directly into power, and soon power needs increase exponentially. In the United States, 22 per cent of the energy converted drives vehicles, and another 10 per cent keeps roads open for them. The amount of energy is comparable to the total energy – except for domestic heating – required for the combined economies of India and China. The energy used up in the United States for the sole purpose of driving vehicles built to accelerate beyond bicycle speeds would suffice to add auxiliary motors to about twenty times that many vehicles for people all over the world who want to move at bicycle speeds and do not or cannot push the pedals because they are sick or old, or because they want to transport a heavy load or move over a great distance, or because they just want to relax. Simply on the basis of equal distribution on a worldwide scale, speeds above those attained by bicycles could be ruled out. It is of course mere fantasy to assume an egalitarian consensus sufficiently strong to accept such a proposal. At closer inspection though, many communities will find that the very same speed limit necessary for equal distribution of mobility is also very close to the optimum velocity giving maximum value to community life. At 20 mph constant speed Phineas Fogg could have made his trip around the world in half of eighty days. Simulation studies would be useful for exploring imaginative policies that seek optimal liberty with convivial power tools. To whose advantage would Calcutta's traffic flow stabilize if speeds were limited to 10 mph? What prie would Peru's military pay for limiting the nation's speed to 20 mph? What gains in equality, activity, health, and freedom would result from limiting all other vehicles to the speed of bicycles and sailing ships?
>
> *Tools for Conviviality* (1973)

There is sometimes a strong element of 'mere fantasy' in Illich's arguments, and he likes to exaggerate a point to drive it home: a forty-day round-the-world cycle tour would put Merckx to shame.

'. . . an egalitarian consensus'. Little Red Books and bicycles in Nanking.

One of his most pleasant calculations destroys the myth that the speed of motor transport actually saves time:

> The typical American male devotes more than 1,600 hours a year to his car. He sits in it while it goes and while it stands idling. He parks it and searches for it. He earns the money to put down on it and to meet the monthly instalments. He works to pay for petrol, tolls, insurance, taxes and tickets. He spends four of his sixteen waking hours on the road or gathering his resources for it. And this figure does not take into account the time consumed by other activities dictated by transport: time spent watching automobile commercials or attending consumer education meetings to improve the quality of the next buy. The model American puts in 1,600 hours to get 7,500 miles: less than five miles per hour. In countries deprived of a transportation industry, people manage to do the same, walking wherever they want to go, and they allocate only three to eight per cent of their society's time budget to traffic instead of 28 per cent. What distinguishes the traffic in rich countries from the traffic in poor countries is not more mileage per hour of lifetime for the majority, but more hours of compulsory consumption of high doses of energy, packaged and unequally distributed by the transportation industry.
>
> *Energy and Equity* (1974)

Even if the figures seem snatched from mid-air, the point is excellent. Many motorists or 'habitual passengers', as he calls them, will surely recognize themselves in this account. Illich's own socio-political jargon is scarcely a 'convivial tool', but his unusual perspectives on the modern world deserve more attention. Anyone interested in fitting the bicycle into an overall revolutionary critique of the relationship between industrial and agrarian societies should read his works for themselves.

Illich's urgent message for the developing nations may, alas, be shouting against the winds of change. The coastal nations of Africa and South America are busily hunting for oil. The oil-rich countries show few signs of spreading the wealth of the ruling classes, let alone of following Illich's advice. Iran is anxious to put up the price of oil as much as it dare so that its economy can be fully industrialized by the time its relatively small supply of oil has run out. On the other hand the largest oil-owners, notably the Saudi Arabians, want to keep prices low so as to avoid frightening present oil-users into

Bicycles and a sewing machine (their industrial predecessor) ride together in the streets of Peking.

finding other forms of energy. Two thirds of the world seems poised indecisively between modern and traditional society; perhaps they are waiting to see how we in the hyper-industrial nations (both communist and capitalist) will emerge from the laborious concatenation of mistakes and problems that afflicts us at present.

What does the future hold for the bicycle? Firstly, an appropriate balance with the motorcar must be established. The economic fabric of European and American society depends on road transport. The working vehicle is here to stay for a long time and even the most inveterate car-haters must admit this. The private motorcar's future cannot be so certain. It would be wrong to deny the fact that the car has created happiness for many millions of people during the first three quarters of this century. But can we still afford the freedom of individual mobility if it has to be exercised entirely through the

'. . . the speed of bicycles and sailing ships'.

automobile? If the car has to have ever more complex and restrictive systems of roads, motorways, traffic controls and taxes to support it, does that original freedom even exist any more? It would be painful to lose our cars, however clear-minded we may be about the excellent arguments that convince us they should go. In the meantime, let the motorcar move over a little to make room for the bike. Those who say they *would* cycle, if it were not for the number of cars, should take heart and try it. Then there will be less traffic. Let transport policies feature the bicycle as an important element in an integrated system. Governments, for their own good and for the good of the population, should advance cycling as much as possible, by planning, by taxation and by road-safety legislation. The industry must continue to refine the utilitarian bicycle and not only the super lightweight. City centres must become quiet, car-free and well served by public transport, even if bikes have to be banned from

them as well. The aims of those who see the two-wheeler as an instrument of athletic prowess and those for whom it is a symbol of ecological wholeness are not as far apart as they may seem.

Whether cyclists see themselves as the spearhead of a social revolution, as sportsmen, tourists, hardy commuters or fair-weather weekend riders, they all share in that original freewheeling thrill of

motion and balance on two wheels. With only a little encouragement the bicycle may yet bring us gently back into tune with the real world of time, weather, physical effort and the gravity that plays on all things under the sun.

Appendix 1

Maintenance, Comfort, and Safety on the Road

MAINTENANCE

The maintenance of your bike is vitally important. Your life depends on it – every time you ride, not just occasionally. And the maintenance is up to you. Common sense dictates that you teach yourself at least the basic elements of cycle repair, however much you may think of yourself as chronically unmechanical or above such things.

This book is not a technical manual. There are plenty of good technical manuals available, covering both the simplest and the most complicated of bike repair jobs. On the basis of a few years' return to cycling, not wholly unsullied by breakages, errors and frustrations, we have collected our reflections on the business of caring for a bicycle, which we hope will persuade new riders to look after and ride their mounts in the right frame of mind.

For routine oilings and adjustments, look to a collection of repair books. Like cooking, in mending a bicycle there are often several different ways of arriving at a satisfactory result. Just as few interested cooks depend on one cookery book, a small bundle of bike maintenance books is better than one. Those with drawings are usually clearer and more immediately comprehensible than the ones filled with photographs. All such books will contain pages of redundant information intended for different kinds of machine, and explanations of parts you do not possess. Unfortunately, many will seem to avoid the particular problem you face, and may even contain misleading or misleadingly partial information. One of the best books known to us omits the essential piece of information that the so-called 'fixed cup' in the bottom-bracket assembly has a left-hand thread. Without this knowledge the enthusiastic home mechanic might bend the whole frame out of joint or strip the threads in the bottom bracket. Of course, you also need to know that Italian frames often have fixed cups with normal right-hand threads. This may all seem esoteric, but the point is that technical manuals are highly necessary but sometimes unreliable. Also they only tell you how to deal with machinery that is in utopian good shape, without any stuck, broken or problematic bits. In fact they usually miss all the genuine difficulties of do-it-yourself repairs – all the mistakes, muddle and bodging that take up the time and cause irritation and rage.

Let us consider the case of a stuck cotter-pin. One day while riding uphill you notice a regular click which you do not recognize; it coincides with every revolution of the cranks and it quickly gets a bit worse. After a cursory examination it is clear that one of the cranks is fractionally loose on the axle, resulting in an annoying click at every thrust on the pedal. As the nut on the cotter-pin that holds the crank to the axle can be tightened without difficulty, there seems to be no problem in taking up the slack. The situation is still under control. You find the appropriate spanner and prepare to complete an instant repair, logical, simple and without snags. To tighten the nut on the pin a certain modest force is needed. The click remains. You pull a bit harder on the spanner, and zilch! – the thread is stripped. Cotter-pin still firmly stuck in place, still with click, but no nut. The bike is as yet rideable, but it needs a new pin. Eventually, you argue, it will work loose, so it must be mended soon.

Repair manuals make the removal of cotter-pins utterly straightforward – a matter of routine, clear-minded, clean-handed maintenance to the bottom-bracket assembly. The process is simple: undo the nut a fraction (yours is shorn away, of course); protect the crank and bottom-bracket shell with appropriately placed pieces of wood; tap the pin free, protecting it too with wood. Bingo.

After precarious experiments with odd pieces of firewood it becomes clear that a proper 'tool' must be made, and so you drill a suitable hole in a large wooden block. On this block the bike can be finely balanced, and your hands are free to perform the other necessary rituals prescribed by the manual. Proud of your initiative and eagerly anticipating success, you give the silly pin a tap with a hammer, not forgetting of course that it should be shielded from your strength with another piece of firewood. Two or three more smartish taps, and then a more purposeful thwack. The pin has not budged a millimetre. After banging away for some time you leave it, hoping it will sort itself out. Perhaps that rain of blows on the pin (you have abandoned the protective wood) has made it heat up infinitesimally and thus expand. A sharp blow when it is not expecting it may be the answer.

No good. You sluice penetrating oil into the gap where the click is and leave it overnight. The pin may come to its senses in the morning. During the night you wonder if the penetrating oil, too liberally applied, will penetrate deep into the bottom bracket and there dissolve grease, causing havoc with the ball bearings, future disasters and more cotter-pin removals, but there is nothing you can do about this anxiety.

Next day a few lusty thwacks (the pin is now sadly bent) do not prevail. Will it have to be drilled out? Is this a case for a professional repair man, a humiliating trip to the mythical 'bike shop' which exists in the pages of American repair books? Will it take a month for the job to be done because of the shop's backlog of work? In a final despairing look at the situation you notice that the hole drilled in your special block of firewood is dented at the bottom with a shiny circular concavity, very much like the imprint of the cotter-pin head. Blockhead! The hole is not deep enough. You have been bashing the pin into the wooden block, without a chance of it coming free. Worse, that block had not been providing any protection to the crank or bracket shell, and, according to the books, the whole caboodle may be knocked awry. Back to work with the drill. Set up everything again. A sharp tap. Nothing. Frustration and rage begin to seep back. A thwack – and behold! The pin is loose, ready to be tapped free; the job is done – or, rather, begun.

The account is still incomplete. It leaves out the number of times you shouted at people to go away; your sulks at not being able to ride the bike just when you wanted; the sense of total bafflement, impotence and frustration. Nor was a new pin easily fitted. For a start, there seemed to be several different sizes and it took two visits before you had bought the right one, and even then three pins were needed to get a perfect fit, despite filing them to shape, banging them in and removing them several times.

Even the simplest repair can become like a maze, a frustrating labyrinth of wrong turnings, blind alleys, multiple choices, possibilities, errors and bungles. It may be that you never have to touch a cotter-pin in your life, or you may find their removal a task of ease and simplicity. But at some stage in your relationship with a bicycle the chances are you will experience what Pirsig in *Zen and the Art of Motorcycle Maintenance* calls 'stuckness'.

Suppose, after your humiliation with the cotter-pin, you fit a cotterless chainset to your bike. One day, after many miles of trouble-free riding, your ears pick up a familiar 'click'. With sinking heart you note that the unwanted noise occurs, just as before, every time the left-hand crank goes past Top Dead Centre. Experience tells you that a cotter-pin is loose, but you have no cotters. What, then? Could the bolt holding the crank on the end of the axle need tightening? You try, but it is firm, and the click remains, sporadically. You start to discriminate between possibilities. When the click occurs, it coincides with the movement of the crank and chainwheel. It could be:

1. Your left knee?
2. A cracked left crank? No such visible, but then you have read of the dangers of metal fatigue in cycle alloys. This is worrying.
3. Bracket bearings breaking up? Why should it occur only with the left crank under pressure?
4. Left pedal nearing retirement? Quite a good idea, but it seems to spin freely, without too much play.
5. Left toe-clip, strap or even shoe hitting the crank? Not obvious, but it does not seem to be any of these either.
6. A badly fitted rivet on the chain hitting the front changer? No: the chain passes freely through the cage on the changer. Also this would not be a regular noise in time with the revolutions on the chainwheel.
7. A bent tooth on the chainwheel hitting the front changer? None visible.

You are stuck. None of the explanations seems to fit the case, but you are thinking along the right lines now. You establish an important point. The offending noise only occurs in high gears, that is to say, when the chain is on the large chainwheel. This makes Explanation 7 seem the most plausible, as the problem seems to be something to do with the front changer, but then, on the other hand, high gears mean more pressure on the pedals and so it could still be 3 or 4. And 2 has not been excluded yet.

By now you will probably have tried out some of these possibilities, bought new pedals or a new crank, or given up. Time passes and the click comes and goes, always in high gear. But many careful inspections have shown that when the left crank is passing Top Dead Centre the chainwheel and chain are moving freely through the changer. One day you determine to go through it all again, carefully. For once you look at your right crank, the one that does *not* click when you put pressure on it. There on its inner surface you notice a slight curved scratch or groove. The solution hits you like a custard pie. Certainly it only clicked when you pushed on the left crank, but only because the *right-hand* crank was sneaking up behind and hitting the *back* of the front changer, not where you had been looking at all. You prove the new hypothesis in thirty seconds. Half a turn's adjustment with a screwdriver to the changer and you can be certain that the click is banished. The reasonable association of the offensive noise with pressure on the left crank had led you to overlook the simple, right-under-the-nose solution to the conundrum. You will never be caught out like that again.

You may not, indeed, but the chances are you will be, at least until you are a very experienced cycle mechanic. However mobile you try to keep your investigations, however alert and close to the cutting edge of thought you may be in making them, and however simple, straightforward and mechanically naked the bicycle's moving parts are, there are always new combinations of possibilities to test you.

Why bother about a click, anyway? If it had been a cracked crank it would have been dangerous. Whether cracked cranks actually click is a moot point. But alloy cranks do break, sometimes suddenly, and at the wrong moment this could send you under a bus, so unless you are cocksure in your metallurgical information it is worth taking a good look at suspect cranks from time to time. Certainly it is foolish to ignore the little warning noises your bike might be sending you, for the early identification and rectification of faults is sure to save time and money.

Maintenance involves a continual series of checks and adjustments. Bikes that are used regularly need regular attention – a relationship rather than sporadic attempts to make good the damage caused by long spells of neglect. Take, for example, the chain. If you spray it weekly with some brand of water-dispersing lubricant (like WD 40) and wipe off excess grit and grime at fortnightly intervals your chain should stay flexible and clean. One big oiling followed by months of neglect will result in a filthy clogged-up chain and rear sprocket, a poorly functioning derailleur and mucky trouser cuffs. Without lubrication of any kind rust will set in and you will soon have to throw the chain away. The choice is yours.

Probably most people feel that they are 'not mechanically minded', and even believe like Max Beerbohm that the bike gratifies 'that instinct common to all stupid people, the instinct to potter with machinery'. Oily chains are not for them, let alone the innards of bottom brackets. Many of us live surrounded by events and objects that we do not understand, thus immunized from curiosity and wonder by sheer habit. Electricity, plumbing and even the most basic household objects are 'dull'. We only *use* them – they are what is called 'labour-saving', after all. It seems a pity that we settle for such a shallow relationship with our manufactured environment. Perhaps the bicycle and all its attendant frailties can provide a small initiative for escape from this hackneyed situation. Its sources of power are not mysterious, but as familiar as your own knees. True, we will not build it from raw materials. But we can at least master its methods of functioning, its mechanics. Start with the theory. Read the repair manuals with their contrary advice. Then read the bicycle itself.

It is only through experience that mechanical skills can be acquired. The best – the only – way of finding out about hubs is to take one apart. Spread newspaper on the floor to catch escaping parts such as ball bearings (surprisingly small) and place each bit in a careful order that will remind you how to put them back together again. Have plenty of rags handy to clean up as you go along. Keep stopping and thinking about what you are doing. Avoid getting in a tizzy or a temper. Take things apart with as much concentration as you are going to need for putting them back together. And when you are building up rather than dismantling, avoid overtightening nuts, especially when working with alloys. Again this is a thing you can only learn through your fingers and not from books. Check the job as you go along and try not to work when you are tired or in a hurry.

It is very easy to have all the logical explanations in mind of, say, how the brake shoes should be fitted and then absent-mindedly put them in facing the wrong way round. Try to go for a short ride to check the adjustments you have made, and remember that bolts may need tightening from time to time until they settle down. If you buy a new bike, it too will probably require servicing and 'tuning', although a reputable dealer *should* have done most of this for you.

Your first efforts are bound to involve any number of minor snags and calamities, but the knowledge gained will be practical and genuine. You will need one or two special tools and a few good spanners to fit the bolts on your machine. Multi-hole 'bicycle spanners' may be all right for roadside repairs but they are not good tools. Common sense, an open mind, a good deal of patience and a readiness to make mistakes are all that is needed. The mistakes are seldom *that* bad.

COMFORT

Having chosen a frame of the right size (see Chapter 2), the main task is to accommodate the bike to the rider. Most people start to cycle with their saddles too low. That way their feet are closer to the ground and so they feel safer, not a bad thing at all for beginners. But they may carry on believing that it should be possible to place the feet on the ground while sitting in the saddle. On an adult bicycle this is not the case. On a properly adjusted bike the serious rider will always have to ease forward off the saddle in order to be able to stand on the ground while straddling the top tube. The precise manner of arriving at this proper adjustment, of establishing how the saddle and handlebars should be placed in relation to the immovable fulcrum of the bottom bracket, is subject to dispute amongst different sorts and different generations of cyclists. Stems, frames, handlebars, cranks and riders all come in different shapes and sizes, so the possibilities are legion.

Cyclists on shallowish frames, with angles of 72 degrees parallel or less, sit quite far back (the tip of the saddle two inches or more

behind a line drawn vertically up from the centre of the bottom bracket). They might practise the art of 'ankling', flexing their ankles in such a way as to help push the cranks through Top Dead Centre. Modern racers, especially time triallists, use steeper frames and longer stems and sit more directly above the bottom bracket.

Many do not 'ankle' in this position, and when pulling hard on the handlebars they finish up even further forward.

Bearing in mind the different purposes for which bikes can be intended, it would be unwise for us to give a cast-iron rule about riding position. Comfort is in any case relative. People who have not cycled for some time will not find a bike as blissful as an armchair, and initial aches and pains may come to be confused with the problems of finding a good riding position. If you are new to the game it will take some time to get comfortable, both to wear you in and to fiddle the bike into just the shape that suits you. Suppose you have bought a medium-quality mass-produced 'racing bike' (not intended for racing) and you have not cycled for some time. Let us start with a rough and ready initial adjustment and then move on to refinements.

The most important measurement is from pedal to saddle. Take your shoes off, find a convenient wall or piece of furniture to lean on and mount your bike. When you are sitting firmly on the saddle you should be able to keep your heel on the pedal at the furthest point from the saddle, that is, when the crank is in line with the seat tube and your leg is straight. (When pedalling with the ball of your foot your leg should be slightly bent at this 'furthest' position.) Raise or lower the saddle height by tiny amounts (less than a quarter of an inch) until this is achieved. Assuming that the saddle is placed sensibly and centrally on the saddle pin, and that your standard bike has an average-sized stem (and that you too are of average build), then you will find that when you place an elbow against the nose of your saddle, your fingertips will be between one and half and two and a half inches from the handlebars. Start with the tops of the bars level with the saddle, but let comfort decide. Ride around for a bit with your bike arranged like that and give your body time to get used to what seems an unnatural position. When you make variations, make them very slight. Even a quarter of an inch change to the saddle position can profoundly alter how the cycle feels under you. Big alterations could surprise and strain muscles. Ride yourself in gently.

For more elaborate adjustments you will need an accurate and unstretchable measure, preferable in centimetres, and, if you get really obsessive, a plumb line. An intimate partner who is prepared to measure your inside leg several times can be more than helpful. In

1965 physiologists studying ergonomics at Loughborough College of Physical Education produced a lengthy research paper on the optimum riding position for cyclists so far as transmitting strength to the pedals was concerned. They concluded that for maximum power the distance from the pedal spindle to the top of the saddle (again with crank in line with the seat tube) should be 109 per cent of the inside-leg measurement taken from crotch-bone to floor in stockinged feet. Read this measurement in centimetres, multiply by 109, shift the decimal point, and adjust the saddle as necessary. Bear in mind that you should make the move to the new position gradually.

As we have already said, saddle position has caused arguments amongst racing cyclists, who are at the forefront of cycling theory. The discoveries of scientists in relation to sport have not been proved correct in every case. When the Loughborough group announced their conclusions some cyclists were astonished at what seemed a startlingly high saddle-height. Some racers still pooh-pooh the 109 per cent theory and do well, especially, perhaps, those who advocate fast pedalling rather than huge windmill gears as a means of race-winning. This last is a particularly well-worn controversy amongst time triallists.

If you have to move your saddle up, you must keep in mind the now altered relationship between the saddle and the handlebars. The best method is to arrange and check your riding position in a carefully ordered sequence. The following adjustments are intended for the keen rider on a bike with drop handlebars:

1. *Vertical saddle adjustment*. Pedal-spindle centre to saddle top (crank in line with seat tube) = 109 per cent of inside leg length.

2. *Saddle tilt*. The saddle should be level, but some riders prefer a slight upward or downward tilt, which let comfort dictate.

3. *Fore-and-aft saddle adjustment*. When the cranks are horizontal the pedal spindle should be vertically below the knee joint of the leading knee (or a point just behind the knee-cap). You may well have to move the saddle forward (and then eventually lengthen the stem) to achieve this, if you are finding that the 109 per cent formula suits you. If your frame is 72 degrees and your thighs are short, you may find that you cannot arrive at this position without having the saddle perched precariously far forward. In this case, lower the saddle a little.

4. *Reach to the handlebars.* The same problem as the last, only at the front of the bike, and you may need to buy a new stem. If you place your elbow against the nose of the saddle, the handlebars should be around two inches from your fingertips. Do not mislead yourself into thinking that a shorter stem will mean more comfort. Curiously enough, if you find yourself sliding forward in your saddle it could be that your stem is too *short* rather than too long.

5. *Height of stem.* A matter of personal preference depending amongst other things on how deep the drops of your bars are, whether you want to race, and so on. If you have not made up your mind how low down and sporting you want to be, then start with the stem the same height as the saddle and move it downwards gradually. Naturally, you should always be able to see the road ahead without getting a crick in the neck, even when racing and sheltering from the wind.

The last two adjustments can be done at any time, as they do not affect the relationship between seat, pedals and handlebars.

6. *Tilt on bars.* This can make a surprising difference to the comfort of your wrists and elbows. Most people seem to prefer a slight upward angle on the lower part of the bars. Try setting the bars so that a line from the hooks would run through the axle of the rear wheel. Others like part of their bars horizontal. Experiment, gradually.

7. *Position of brake levers.* Important, because you need to be able to ride in comfort holding the brake hoods. You should be able to operate the brakes from this position, as well as from the bottom part of the bar. As a guide, the levers should be attached (very firmly) so as to point vertically downwards, in so far as this is possible given their hooked shape.

SAFETY ON THE ROAD

When you are riding a bike, the number-one enemy to safety is yourself. Your own inattention, stupidity, arrogance, vanity or folly are all liable to kill you; and you can kill yourself on a bicycle without any help from the motorcar, just by falling off and breaking your head on a kerb. To ride safely you need that same ever-attentive common sense that is indispensable in the business of looking after your bike. You can and should look at all relevant material on road safety, the highway codes that apply wherever you may be riding. You need to know the law. But no book-learning will ever help you unless you start out conscious of your responsibility to yourself and to others. Highway codes are often vague about the special problems that face cyclists. A particular bugbear for British cyclists, for example, is the urban roundabout. The guidelines in the *Highway Code* on how to negotiate this jolly whirligig of speeding cars and lorries may be sensible enough for the invulnerable motorist, but they are more or less useless for the cyclist who actually has to risk his life in lanes of dense and fast-moving traffic. What can the cyclist do in city riding to make himself a less likely victim in the brutal stampede home from work? How can he be more defensive?

First of all he can be as conspicuous as possible, wearing brightly coloured clothing (white or yellow rather than camouflage browns and greens), and this is particularly important at night. He must ride positively and predictably in his proper space on the road, rather than darting in and out of the gutter to avoid parked cars, drain-holes or rough bits of tarmac. Stay as visible and predictable as possible at all times.

Secondly, a cyclist should be in perfect control of a well-kept machine. He should be able to glance behind him without difficulty or discomfort, making good use of his 360-degree vision with no blind spots. Avoid clothing, like hooded jackets, that impairs your vision – or your hearing, that other essential aid to knowing the state of the road behind you. Changes of direction, hand signals, and so on, should be performed with confidence and in complete knowledge of the road conditions both before and behind. Though it is foolish to make a habit of it, a cyclist ought to be able to ride without holding the handlebars, as a test of his own skill and relaxation and as a useful way of being sure that the alignment of his frame and wheels is spot-on. If you do not feel relaxed, confident and competent on your bike, then stay away from all busy roads until you do.

Thirdly, like the motorist, the cyclist should be constantly thinking ahead, anticipating dangers to be avoided rather than waiting to see what happens. If someone thoughtlessly opens a car door and knocks you off (how many times have you nearly done that to a passing cyclist?), in most circumstances it may well be your fault for riding too close to a parked vehicle filled with unpredictable passengers. Does that seem hard? So is the edge of a car door. Experience will increase and refine your ability to foresee possible dangers. For example, when you are overtaken by a car, beware of assuming that the road behind is now clear. You heard the first car coming up and passing, but it is the second one following closely behind that will get you if you are not careful.

Clearly none of these precautions will prevent a drunken madman in his Mustang–Cougar–Jaguar–Tiger GT from mowing you down at ninety miles an hour. You may have been fully visible for half a mile ahead – only twenty seconds to him. The criminal lunacies of a few motorists affect not just the bike-rider but wiser drivers and the pedestrian too. When you are on your bike, however, *all* cars present a constant *possible* danger. The cyclist is at a natural disadvantage when forced to share space with half a ton of dirigible metal, and from the car's point of view a man on a bicycle is weak and vulnerable – an evolutionary and even a social inferior. Beep-Beep! The automobile turns the nicest folk into Toad of Toad Hall.

Cars do some very odd things indeed when they find themselves momentarily inconvenienced by a bicycle. Perhaps as more people go back to self-propulsion, manners and habits will improve. But at present a car-watcher on a bike can observe strange and terrifying examples of road behaviour.

The Slow Reactor. On a perfectly clear and straight road a car will roar past from behind, only inches from your elbow, and then, ten yards ahead, it will suddenly swerve into the middle of the road with an indignant hoot to show retrospective annoyance at having to depart from its intended flight path.

Overtaker's Grimace. Two cars are racing towards you, one overtaking the other, and, as far as you are concerned, the one in the middle has slightly misjudged his distances. As you ride for the ditch you will see a fixed grin on the overtaker's face. He is actually more frightened than you, but nothing would make him give in and slow up. Motorists indulge a curious bigotry where the cycle is concerned. They will decelerate meekly for a large load, or a combine harvester. It has got four wheels, after all, even if it is going at 5 mph. But to have to slow to 15 mph for a mere two-wheeler seems to set their blood boiling and they may start to behave dangerously. Drivers usually avoid eye contact when they get into this mood. They will glance away, address their wives or peer in the glove box rather than look at you.

The Nervous Passer. Usually found on winding country roads. Some drivers seem to have inflated notions of the size of their own vehicles, and they will crawl along behind a cyclist, occasionally hooting impatiently, when there is in fact plenty of room to pass.

These are only the commoner behavioural oddities.

The Cyclists' Touring Club provides free third-party insurance and legal aid for its members, and every two months they publish in *Cycletouring* how much has been paid out in damages under this scheme. Of course, this report only shows successful claims against motorists and local authorities and not the cases where it was the cyclist at fault. Nevertheless, the figures are revealing because they indicate the kind of accidents which can maim even the most careful rider. The list for 1975 totalled over £28,600 in compensation involving 113 cases with cars. The most common incidents were as follows:

Cars turning in front of cyclists	28
Cars overtaking	26
Cars emerging from side roads	20
Bad road surface	16
Car doors opening	7
Cars at roundabouts	4
Dog	1

The list speaks for itself, and may well persuade you to join the CTC or some similar large organization.

The following photographs come from a campaign mounted by the Royal Society for the Prevention of Accidents to 'Give Cyclists a Chance'. We use them, with our own comments, to illustrate the most common dangers for riders and drivers alike.

1: A bike can travel nearly as fast as a car in town, even faster if the traffic is heavy. Motorists should remember this if they meet a cyclist near to junctions or side roads where they have to turn – don't cut him off.

2: Overtaking drivers should give bikes a wide berth. The slipstream from a truck can make a rider wobble or even suck him further into the roadway. Motorists seem to find it difficult to judge the speed of a bike, so watch out for the car that moves out too late or cuts in too soon.

Riders should take care if they find themselves between the kerb and a truck or trailer at junctions or roundabouts. The driver may not see you back there and the tail-end of his outfit can knock you off if he clips the corner as he turns.

3: Drivers may see you but still not 'register' your presence and speed. Consider the car that edges over the white line slowly. The driver wouldn't *crash* into you, but he thinks that creeping into you is O.K.

Watch out for the car joining a fast carriageway from a slip road you have already passed. The driver will often be looking over his shoulder at following traffic while he moves up behind you.

4: Bikes have to cope with a variety of bad surfaces and obstacles that cars can ignore. Motorists should try to sympathize and give them enough room to take a sensible and steady line a few feet away from the gutter. Riders soon learn to watch out for pot-holes, raised drain covers, loose rocks, etc., especially when cycling at night.

5: This picture speaks for itself: riders beware of parked cars with people in them. They may also pull away without checking behind.

6: Cyclists are in danger at roundabouts or right turns across the flow of cars, and so much depends on traffic conditions and the individual skill of the rider. Drivers should try to be patient and riders should try to be steady and predictable, giving bold signals and checking to be sure that they are 'received'. Establish eye contact if possible: glare at each other by all means (it's better than colliding) but politeness is best.

On the other side of the coin there is a kind of myth that the cyclist himself is always as gentle, polite and unaggressive as his machine. Self-righteousness and holier-than-thou judgements are unattractive, even if they are born out of honest toil and the ecological soundness of self-propulsion. Alas, cyclists can behave like lunatics in a whole variety of inconceivably daft ways, taking risks just for excitement, perhaps. And they can be as aggressive as anyone, picking fights with trucks or shouting at people crossing the road with all the selfish Darwinism that they would chide in the motorist. It seems that machines for going faster than a walking pace bring out the worst in man, and the bike is no exception. Perhaps things will get better as transport priorities change. All we can recommend is vigilance, calm and common sense for all road-users, be they walkers, riders or drivers.

GREAT SELF-RESTRAINT.

Lady in Pony-cart (*who has made several unsuccessful attempts to pass persevering beginner occupying the whole road*). "UNLESS YOU SOON FALL OFF, SIR, I'M AFRAID I SHALL MISS MY TRAIN!"

Lady in Pony-cart (*who has made several unsuccessful attempts to pass persevering beginner occupying the whole road*): 'Unless you soon fall off, Sir, I'm afraid I shall miss my train!'

Punch, 1896

Appendix 2

Ordering a Hand-built Frame

These notes are intended as a guide to buying a custom-built frame. They try to lay out the choices you will have to make in something like their order of importance.

The first thing to say is that there are many good frames available off-the-peg and they offer a choice of sizes and fitments to suit most people. Do consider them, particularly if price is an issue. A tailor-made machine will cost a little more and take a lot longer to arrive. (Always treat a builder's delivery date with a pinch of salt; then you could be pleasantly surprised.) Develop an 'eye' for detail on a bike and take every opportunity to examine other frames closely. Gradually you will be able to detect small differences in design, manufacture and finish between one frame and another, even from the same builder. Once you have this 'eye' you will be able to select what you want on a stock frame with more confidence or check over one that has been built for you. There are many decisions to be made when you order a custom frame, but they all come back to what sort of riding you want to do.

PURPOSE

Is your bicycle for touring or general riding, racing or time trialling? If you want a compromise between these functions then a machine with relaxed road-racing lines might be the best bet. Remember, however, that a purpose-built touring frame can offer unrivalled convenience and comfort, especially if you want to carry even a little luggage. Do not be trapped into thinking that the only bicycle is a racing bicycle. Tell the builder what you want, anyway, and listen carefully to his advice.

MATERIALS

Many good bikes have plain-gauge tubing, but if you are going to the expense of having one built for you, then 531 double-butted is sensible; the extra cost of the materials is such a small part of the overall price. 531 Super Light is intended for moderate road racing and use in time trials. Reynolds 753 is expensive, but its particular combination of strength and lightness recommends it to the racing fraternity. Note that it is a demanding material and the builder will not consider repairing it if the frame is damaged. 753 comes in different gauges according to its purpose and your weight. Columbus, Vitus and one or two other firms also make excellent lightweight tubing, and it may be that your builder will have his own preferences in this regard.

BASIC DESIGN

The first points to establish are the size of the frame and whether you want clearance for high-pressure tyres, tubulars or mudguards. Unless it is to be an outright racing machine, be sensible and consider mudguards. Why get soaked in the rain? If you want an open-frame woman's bike, the 'mixte' style is generally held to be the most rigid. But many women prefer the strength of the diamond frame.

ANGLES

Are the frame angles to be on touring lines at 72 degrees parallel, or steeper at 73 degrees or more? If you are a novice, do think carefully before you order angles over 73 degrees. If you are smaller than average or have short thighs, it may well be that you will benefit from a seat-tube angle of 74 degrees or more. Another way of expressing this is to note the distance between the top of the seat tube and a point on the top tube vertically above the bottom bracket. The smaller this measurement is, the steeper the seat tube. Continental experts prefer this to 'angles'. Remember that a steep head-tube means more 'nervous' handling. Again, ask your builder's advice.

DIMENSIONS

Ask again about fork rake and the length of the chainstays. Short chainstays give a stiffer, more efficient transmission line to the back wheel and improve the bike's sprinting and hill-climbing characteristics. Racing machines might have $1\frac{1}{2}$-inch fork rake and $16\frac{1}{2}$-inch rear triangle or less. The rear triangle, (rear centres) is the distance measured from the centre of the bottom-bracket spindle to the centre of the rear axle. The front centres measure to the front axle, and unless it is $23\frac{1}{2}$ inches or over you may have trouble with the pedals overlapping the front wheel and hitting it when you turn sharply. If the chainstays are too short there may be problems with chain alignment when changing gear, especially if you use widely spaced ratios. Thus a touring bike will have rear centres measuring $17\frac{3}{4}$ inches or more. The overall wheel base of an average touring machine might be 42 inches (with a $2\frac{1}{4}$-inch fork rake), while a racer might measure $39\frac{1}{4}$ inches. On the subject of chainstays, it is vital to specify the size of chainwheel you intend to fit. The appropriate chainstay has to be dimpled in the right place to ensure clearance.

Modern club or racing frames are 'square', that is, the length of the top tube is the same as, or even less than, the length of the seat tube. The length of the top tube can vary according to how tall you are or how long your back is, and changes here may mean changes in the angles too. In spite of these complications most people can be accommodated. Time trial and criterium bikes often have a high bottom bracket – some $10\frac{3}{4}$ inches off the ground – to allow for hard pedalling through corners. A long-distance racer or touring machine gains a little in handling and comfort by being set slightly lower at $10\frac{1}{4}$ inches.

It will help your builder at this stage if you can also tell him what make and type of wheel hubs, gears, brakes and seat pillar you intend to use. Even more important, tell him what bottom-bracket set and headset you want and get him to fit them as well.

Now we come to the details and fittings.

DROP-OUTS

It makes sense to fit only forged drop-outs to a quality frame, and Campagnolo, Simplex, Shimano, Huret and Zeus all make good ones. The rear drop-outs will have an integral hanger for the gear mechanism, and built-in adjusting screws make sure that the rear wheel slots into centre line accurately and quickly every time.

'Vertical' or 'short' rear drop-outs are in fashion, but they are only essential if you want a very short and tight rear triangle and quick wheel-changes. For touring, or for a bike that will be using different wheels or tyres (sprints or H Ps, for example), the longer-style drop-out gives more room for adjustment and is a much better choice.

LUGS AND THE BOTTOM BRACKET

Your builder may already prefer to use a certain kind of lug, or he may give you the choice between plain or curly. Good names include Prugnat, Nervex, Vagner, Bocama and Haden. Cast or forged lugs and bottom brackets are very accurately and rigidly made to the required angles, but they are slightly heavier than pressed steel. The bottom bracket is best left without lightening slots or holes unless for a pure racing bike. These slots make it essential to fit a sleeve to protect the bearings. Such a sleeve is a good idea anyway to guard against the water and scale that can collect in the tubes.

SEAT CLUSTER

The seat cluster offers several styles from wrap-over to shot-in seat stays, with or without allen key bolts at the seat pillar. Here is where an 'eye' for other bikes will help you choose, for everyone has their own preferences. For touring purposes , however, one should go for strength and simplicity, and the shot-in style is perhaps best avoided because it requires skilful brazing directly onto the seat tube. You can also choose at this point whether you want $\frac{1}{2}$, $\frac{9}{16}$ or $\frac{5}{8}$ inch section seat stays. The 'thicker' stays are supposed to be more rigid. If you are going to use centre-pull brakes you can have a special cable bridge brazed onto the seat stays below the rear of the saddle.

FORK CROWN

The best fork crowns are forged steel. The Italian sloping or semi-sloping style is a beautiful piece of work, but slightly heavier than some other designs. Track crowns are strong and light as well, but designed for round fork blades. Some crowns have reinforcing tangs fitted to them, running down the inside of the fork for an inch or more. Continental-style oval fork blades are currently in favour with road men.

FITTINGS

The most common brazed-on fittings are a button to stop the gear-lever clip from sliding on the down tube and a cable stop for the gears on the right-hand chainstay near the back axle. In addition you can specify tunnels on top of the bottom bracket for the gear cables and cable stops or rings on the top tube for the brake cables. Different bosses can be fitted to the down tube for attaching the gear levers directly and for holding a bottle-cage. Pump pegs can be fitted almost anywhere to choice, or not at all if a cup or a frame-fitting pump is used. Mafac cantilever brakes require brazing directly to the forks and seat stays. The tourist might want special fittings for lamp brackets, dynamo and pannier-carriers. Brazed-on fittings seem sensible, although they are a matter of personal choice. The builder might advise you against some of them if you want to use very light-gauge tubing for the frame.

FINISH AND EQUIPMENT

You will, of course, have your own ideas about the decals, colours and chrome to be put on your bike. Plain enamels last better than 'lustre' or 'flamboyant' finishes, but if you have your frame re-sprayed every two years or so this is not a crucial issue. As for the equipment you want to fit, that is a big question and there is no end to the debate. Your priorities should be a good frame and then a good pair of wheels. You can ring the changes on the rest as you like. Even if you want to race cheaply you should still buy good tubulars and quality racing shorts and shoes. These are much more important than the lightest and latest brake levers or drilled-out chains. Finally, if you are planning your first tour a saddle which suits you is worth more than words can say.

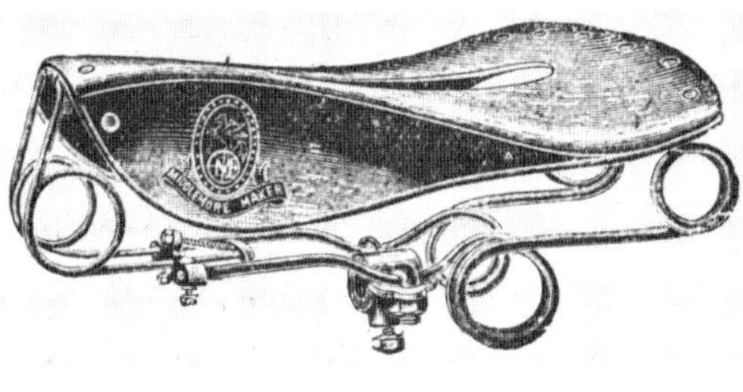

CYCLING
WINTER RIDING.
BETTER THAN MUGGING IN A TOWN ANYWAY!
1930

Books Consulted and Further Reading

We have specified paperback editions whenever possible

MAINTENANCE AND ROAD SAFETY

RICHARD BALLANTINE, *Richard's Bicycle Book*, Pan, 1975. One of the best maintenance manuals, full of enthusiasm for cycling, with some theory and advice on riding and purchasing a bike as well.

Halfords' Guide to Bicycle Maintenance and Repair, 1974. Simple instructions, well illustrated with drawings.

Reader's Digest Guide to Bicycle Repairs, 1974. As above.

PETER ROBERTS, *Better Cycling*, Kaye & Ward, 1972. A well-illustrated book on basic maintenance, road safety and riding technique. Written in collaboration with the Royal Society for the Prevention of Accidents (RoSPA) and directed at young cyclists.

ROSPA, *Skilful Cycling*, RoSPA, Royal Oak Centre, Brighton Road, Purley, Surrey CR2 2UR. An inexpensive booklet on roadcraft, technique and basic maintenance. A 'must' for all parents and young riders.

Bicycles. All about Them, Puffin, 1976. A brief introduction for children, including how to fix up an old bike.

REG SHAW, *Teach Yourself Cycling*, English University Press, 1963. In the 'Teach Yourself' series, it includes history and general information as well.

HISTORY

FREDERICK ALDERSON, *Bicycling: A History*, David & Charles, 1972. A comprehensive technical and social survey.

T. E. CROWLEY, *Discovering Old Bicycles*, Shire Publications, 1973. A little booklet in the 'Discovery' series, very authoritative and excellent value.

ANDREW RITCHIE, *King of the Road*, Wildwood House, 1975. Detailed social and technical history, profusely illustrated and widely researched with a good bibliography.

ROBERT A. SMITH, *The Social History of the Bicycle*, American Heritage Press (McGraw-Hill), New York, 1972. The early life and times of the bike in the States, most entertaining and full of original research.

JOHN WOODFORDE, *The Story of the Bicycle*, Routledge & Kegan Paul, 1970. Contains illuminating reflections on the bike in society.

GREGORY HOUSTON BOWDEN, *The Story of the Raleigh Cycle*, W. H. Allen, 1975. A comprehensive history of the firm, illustrated.

GEORGE A'GREEN, *This Great Club of Ours*, The Cyclists' Touring Club, 69 Meadrow, Godalming, Surrey, GU7 3HS. The story of the CTC, illustrated with photographs and drawings.

C. F. CAUNTER, *History and Development of Cycles*, HMSO, 1958.

ARTHUR DU CROS, *Wheels of Fortune*, Chapman & Hall, 1938. The du Cros family and the rise of Dunlop.

GEOFFREY WILLIAMSON, *Wheels Within Wheels*, Bles, 1966. The story of the Starleys of Coventry.

H. W. BARTLEET, *Bartleet's Bicycle Book*, J. Burrow & Co., 1931. Fascinating accounts of the early riders and their machines, from the founder of the Bartleet Collection in the Herbert Art Gallery and Museum, Coventry.

MARTIN CAIDIN and JAY BARBREE, *Bicycles in War*, Hawthorn Books Inc., New York, 1974. Serious study of its subject with good illustrations.

A. B. DEMAUS, *Victorian and Edwardian Cycling and Motoring from Old Photographs*, Batsford, 1977. Contains some striking bicycle photographs, but is mostly devoted to motoring.

G. A. TOBIN, 'The Bicycle Boom of the 1890s', *Journal of Popular Culture*, Vol. 7, no. 4, 1974.

A. E. HARRISON, 'The Competitiveness of the British Cycle Industry, 1890–1914', *Economic History Review*, XXII, 2, 1969.

EUGEN WEBER, 'Gymnastics and Sports in *Fin de Siècle* France: Opium of the Classes?', *American Historical Review*, Vol. 76, 1, February 1971.

J. A. R. PIMLOTT, *Recreations*, Studio Vista, 1968. A visual history of recreation in Britain.

The Velocipede: Its History and How to Use It, 1869. A CTC facsimile reprint.

JACK RENNERT, *A Hundred Years of Bicycle Posters*, Harper & Row, New York, 1973. A good collection of posters, many in colour, with excellently detailed and informative introduction and notes.

SPORT

GEOFFREY NICHOLSON, *The Great Bike Race*, Hodder & Stoughton, 1977. An entertaining and detailed eye-witness account of the 1976 Tour de France, with many side-lights on its history, organization and personalities. Written by the Sports Editor of the *Observer*, it is of interest to both the enthusiast and the general reader.

PETER WARD, *King of Sports: A Text Book on Cycle Road Racing*, Kennedy Bros. (Publishing) Ltd, Silsden, Keighley, Yorks. Comprehensive instructions and details on how to go cycle racing.

NOEL G. HENDERSON, *Continental Cycle Racing*, Pelham Books, 1970. The classics and the tours described and recorded.

JAMES C. MCCULLAGH (ed.), *American Bicycle Racing*, Rodale Press Inc., Emmaus, Pennsylvania, 1976. The American scene past and present with some insights on Europe.

TOMMY SIMPSON, *Cycling Is My Life*, Stanley Paul, 1966. The autobiography of Simpson up to the year before his tragic death on Mont Ventoux.

REG HARRIS, with G. H. BOWDEN, *Two Wheels to the Top*, W. H. Allen, 1976. Autobiography.

MARSHALL W. TAYLOR, *The Fastest Bicycle Rider in the World*, facsimile edition in Black Heritage Library Collection, Books for Libraries Inc., Plainview, New York, 1974. Autobiography.

PETER CLIFFORD, *The Tour de France*, Stanley Paul, 1965. A lively and detailed account of all the Tours up to 1964; the most informative book available in English.

DAVID SAUNDERS, *Tour de France 1975*, Kennedy Bros. Publishing, 1975. The publishers of *International Cycle Sport* produce these excellent accounts of the Tour and the Giro each year. Illustrated.

J. B. WADLEY, *My Nineteenth Tour de France*, J. B. Wadley (Publications) Ltd, 35 Atwood Avenue, Kew, Surrey, 1974. A most entertaining account of a holiday following the Tour by bike, full of information and colour.

J. B. WADLEY, *Old Roads and New*, J. B. Wadley Publications, 1972. Includes an account of the Paris–Brest–Paris of 1971 in which the author, a leading cycle journalist, took part.

PETER KNOTTLEY, *Cycle Touring in Europe*, Constable, 1975. A guide to bike holidays in Britain and abroad with sections on equipment, routes, etc.

LES WOODLAND, *Cycle Racing: Training to Win*, Pelham Books, 1975.

ROAD TIME TRIALS COUNCIL, *Elementary Training for Cycling Time Trials*, from the RTTC National Coach, R. W. E. Poole, 35 Church Road, Woodley, Reading, Berks. RG5 4QN. Excellent advice on training, equipment and racing in time trials. A valuable little booklet.

TECHNOLOGY AND ECOLOGY

F. R. WHITT and D. G. WILSON, *Bicycling Science*, MIT Press, Cambridge, Massachusetts, 1974. A scientific study of the ergonomics and mechanics of the two-wheeler.

S. S. WILSON, 'Bicycle Technology', *Scientific American*, March 1973. Influential and useful article by an Oxford don.

R. E. WILLIAMS, 'De Motu Urbanorum', *British Medical Journal*, 4 October 1975.

JOE KOSSAK, *Bicycle Frames: A Close-up Look*, World Publications, Mountain View, California, 1975. One of the *Bike World* book series, a unique and comprehensive account of frame-building and design explained for the layman.

ROBERT WRIGHT, *Building Bicycle Wheels*, World Publications, Mountain View, California, 1977. As above, for wheels; also detailed encouragement on how to lace and true your own wheels.

RICHARD FEILDEN, *Give Way: A Report on Cycling in Britain*, Friends of the Earth Ltd, 9 Poland St, London W I V 3 D G, 1975. Essential reading if you want to persuade your local authority to do more for cyclists.

JUDITH GLADINGS, *Alternative Transportation Modes: Bikeway Planning and Design*, Council of Planning Librarians, Exchange Bibliography 941, December 1975 (Box 229, Monticello, Illinois 61856). A bibliography of projects and ideas for the planner.

Cyclateral Thinking, Urban Bikeway Design Collaborative, W 20–002 MIT, Cambridge, Massachusetts 02139. An annual 'atlas of ideas for bicycle planning'. UBDC also produce a comic book on road safety, *Sprocket Man*.

TERENCE BENDIXSON, *Instead of Cars*, Pelican, 1977. A serious and eloquently persuasive study of transport, town planning and the dangers of the private car. One of the best of its kind.

IVAN ILLICH, *Tools for Conviviality*, Fontana, 1975. A formidable criticism of what mass production has done to human life and freedom.

IVAN ILLICH, *Energy and Equity*, Calder & Boyars, 1974. An iconoclastic essay on the Western world's over-consumption of energy and on how we are corrupted and manipulated by our dependence on fast personal transport.

LEWIS MUMFORD, *Art and Technics*, Columbia University Press, 1968. Civilized and illuminating essays on tools, machinery and 'art'.

VICTOR PAPANEK, *Design for the Real World*, Paladin, 1974. A gifted designer and teacher explains how we must rearrange our priorities and forgo 'styling' and restrictive patents. A fascinating book.

DAVID PYE, *The Nature and Art of Workmanship*, Studio Vista, 1971. Reflections on 'quality', 'truth to material', etc.

HERBERT READ, *Art and Industry*, Faber, 1966. An examination of industrial design, first published in 1934.

M. KRANZBERG and C. W. PURSELL, JR (eds.), *Technology in Western Civilization*, 2 vols. OUP, 1967. A massive survey from the stone axe to Saturn rockets.

W. H. G. ARMYTAGE, *A Social History of Engineering*, Faber, 1966.

E. F. SCHUMACHER, *Small is Beautiful*, Abacus, 1974. A most stimulating and humane 'study of economics as if people mattered'. Schumacher has been very influential in the development of 'alternative technology'.

GENERAL

J. DURRY and J. B. WADLEY, *The Guinness Guide to Bicycling*, Guinness Superlatives, 1977. An informative and brilliantly illustrated account of all aspects of cycling, from a glimpse of the past to the present day. Translated from the French, it takes an interestingly Continental point of view on various subjects, especially cycle sport.

R. C. SHAW (ed.), *The Raleigh Book of Cycling*, Peter Davies, 1975. Contains articles on buying a bike, maintenance, roadmanship, sport and touring, etc., also a most useful list of addresses for cycling and touring organizations around the world.

FRED DELONG, *DeLong's Guide to Bicycles and Bicycling*, Chilton Book Co., Radnor, Pennsylvania, 1974. A comprehensive and very detailed breakdown of the technical, physical and social aspects of cycling. A large book full of information.

ROGER ST PIERRE, *The Book of the Bicycle*, Ward Lock, 1973. An excellent general introduction, lavishly illustrated.

SEAMUS MCGONAGLE, *The Bicycle in Life, Love, War and Literature*, Pelham Books, 1968. An entertaining and fast-moving book drawing on literature and history alike.

J. B. WADLEY, *Cycling*, Macmillan, 1975. An introduction in the Leisureguides series.

JAMES WAGENWOORD, *Bikes and Riders*, Van Nostrand, New York, 1972. Well illustrated, with many details of the modern American scene.

DANIEL BEHRMAN, *The Man Who Loved Bicycles*, Harper Magazine Press, New York, 1973. A serious and comic celebration of the two-wheeler and a sustained diatribe against the motor car.

FICTION

H. G. WELLS, *The Wheels of Chance*, Dent, 1964. First published in 1896, a delightful vision of cycling in the 'golden years'.

H. G. WELLS, 'The Land Ironclads', in *Selected Short Stories*, Penguin, 1977.

JEROME K. JEROME, *Three Men on the Bummel* (with *Three Men in a Boat*), Dent, 1957. Comedy on a cycle tour of Germany; not quite the equal of *Three Men in a Boat*, but it deserves to be better known.

RALPH HURNE, *The Yellow Jersey*, Pan, 1975. The hero makes a come-back in the Tour de France. The descriptions of cycle racing are among the best written in English fiction.

STEWART PARKER, *Spokesong: or The Common Wheel*. A musical play, highly praised by the critics in 1976, but as yet unpublished.

FRANK DICKENS, *The Great Boffo*, Pan, 1974. A short tale for four-year-olds, with pictures.

ROBERT M. PIRSIG, *Zen and the Art of Motorcycle Maintenance*, Corgi, 1976. A philosophical 'inquiry into values' in the form of a moving novel.

PERIODICALS

Cycling, Surrey House, 1 Throwley Way, Sutton, Surrey, SM1 4QQ. A weekly paper devoted mostly to the sporting scene in Britain, with frequent articles on equipment, clothing, training, etc.

International Cycle Sport, Kennedy Bros. (Publishing) Ltd, Howden Road, Silsden, Keighley, West Yorkshire. A glossy monthly with colour photographs and coverage of European events and occasional races in Britain, New Zealand, South Africa, America, etc. The publishers also produce many small booklets on famous riders.

Cycletouring, 69 Meadrow, Godalming, Surrey, GU7 3HS. A bi-monthly produced by the CTC, free to their members. Contains much information on touring, equipment and the general status of the bicycle, with informative correspondence pages.

Bike World, World Publications, PO Box 366, Mountain View, California 94040. A well-produced monthly with a bias towards sport and medicine and a liking for first-hand accounts of the cycling experience. The publishers also produce a series of little 'Bike Books' on various topics, some of which are very good.

Bicycling!, PO Box 4450, San Rafael, California 94903. A well-filled glossy monthly with articles on touring and racing and a bias towards technical topics, hints, 'road tests', etc.

Miroir du Cyclisme, Editions Miroir Sprint, SARL, 10 Rue des Pyramides, 75001 Paris. French monthly with magnificent photographs and inside details of the world of professional cycle sport.

Useful Addresses

THE BRITISH CYCLING FEDERATION, 70 Brompton Road, London SW3 1EN. The BCF is the national organization which governs road and track competition. Most members belong to it through their cycling clubs. It provides insurance cover and a touring service as well. You have to belong to it and hold a current licence if you want to race. Under the BCF cycle clubs are grouped into various 'Centres' throughout the country to help organize and oversee events in that area. The Scottish Cyclists' Union and the Northern Ireland Cycling Federation are affiliated to the BCF.

THE CYCLISTS' TOURING CLUB, Cotterell House, 69 Meadow, Godalming, Surrey GU7 3HS. The CTC is the main cycle-touring organization, providing insurance cover, legal aid and much useful advice on routes, riding abroad, etc. It has its own sporting competitions and a regular magazine, *Cycletouring*. The CTC is a national body devolved into District Associations which handle local events and club runs.

If you ride a bike a lot, even just for commuting, it makes sense to belong to one of these bodies. The third party insurance cover and the legal aid they offer are well worth the membership. The CTC is particularly ready to speak up for cyclists' rights and to help you with any inquiries you may have.

Illustrations and Acknowledgements

CHAPTER 3

CHAPTER 4

CHAPTER 5

CHAPTER 6

CHAPTER 7

APPENDIX 1

APPENDIX 2

Index

Numbers printed in *italic type* refer to illustrations